# Data Lake Analytics on Microsoft Azure

## A Practitioner's Guide to Big Data Engineering

Harsh Chawla
Pankaj Khattar

*Foreword by Sandeep J Alur*

Apress®

# Data Lake Analytics on Microsoft Azure: A Practitioner's Guide to Big Data Engineering

Harsh Chawla
Bengaluru, India

Pankaj Khattar
Delhi, India

ISBN-13 (pbk): 978-1-4842-6251-1
https://doi.org/10.1007/978-1-4842-6252-8

ISBN-13 (electronic): 978-1-4842-6252-8

Managing Director, Apress Media LLC: Welmoed Spahr
Acquisitions Editor: Smriti Srivastava
Development Editor: Laura Berendson
Coordinating Editor: Shrikant Vishwakarma

Cover designed by eStudioCalamar

Cover image designed by Pexels

Distributed to the book trade worldwide by Springer Science+Business Media New York, 233 Spring Street, 6th Floor, New York, NY 10013. Phone 1-800-SPRINGER, fax (201) 348-4505, e-mail orders-ny@springer-sbm.com, or visit www.springeronline.com. Apress Media, LLC is a California LLC and the sole member (owner) is Springer Science + Business Media Finance Inc (SSBM Finance Inc). SSBM Finance Inc is a **Delaware** corporation.

For information on translations, please e-mail booktranslations@springernature.com; for reprint, paperback, or audio rights, please e-mail bookpermissions@springernature.com.

Apress titles may be purchased in bulk for academic, corporate, or promotional use. eBook versions and licenses are also available for most titles. For more information, reference our Print and eBook Bulk Sales web page at www.apress.com/bulk-sales.

Any source code or other supplementary material referenced by the author in this book is available to readers on GitHub via the book's product page, located at www.apress.com/978-1-4842-6251-1. For more detailed information, please visit www.apress.com/source-code.

Printed on acid-free paper

*Dedicated to my late grandfather, Sh. Inderjit Chawla*

*— Harsh Chawla*

*Dedicated to my Loving Parents,*
*Affectionate Wife & Endearing Daughters*

*— Pankaj Khattar*

# Table of Contents

# About the Authors

**Harsh Chawla** has been working on data platform technologies for the last 14 years. He has done various roles in Microsoft in the last 12 years. He currently works as an Azure specialist for data and AI technologies, and helps large IT enterprises build modern data warehouses, advanced analytics, and AI solutions on Microsoft Azure. He has been a community speaker and blogger on data platform technologies.

**Pankaj Khattar** is a seasoned software architect with over 14 years of experience in design and development of big data, machine learning, and AI-based products. He currently works with Microsoft on the Azure platform as a Senior Cloud Solution Architect for data and AI technologies. He also possesses extensive industry experience in the field of building scalable multitier distributed applications and client/server-based development.

You can connect with him on LinkedIn at `www.linkedin.com/in/pankaj-khattar/`.

# About the Technical Reviewer

 As a Technology Leader at Microsoft India, **Sandeep Alur** heads Microsoft Technology Center, an Experience Center focusing on helping customers accelerate their digital transformation journey. In addition, he is one of the leads driving Microsoft's AI Country Plan charter, with a key focus on "closing the skills gap and to enhance employability" in the ecosystem. He has donned various technology leadership roles at Microsoft while he managed technical evangelism, ISV, and MSP charters for India.

Sandeep has over 22 years of industry experience providing technology and architectural guidance. His experience ranges from dotcom days to distributed/cloud computing to the next-generation intelligent computing era. He has architected solutions across technology/platform landscapes and is a stern believer of his own philosophy, which says *"Enterprise architecture is an evolution and it grows with innovation."* From an innovation standpoint, he believes that the next big leap is in data science, and AI will elevate customer experience to greater heights.

Sandeep holds a bachelor's degree in Mechanical Engineering from RV College of Engineering, Bengaluru and a postgraduate degree in Data Science and AI from the Indian School of Business, Hyderabad.

You can follow him on LinkedIn and Twitter.

LinkedIn-> `www.linkedin.com/in/sandeepalur/`

Twitter-> `https://twitter.com/saalur`

# Foreword

The current decade marks the beginning of emergence of data that is unprecedented. Aptly so, as we read excerpts from leading technologists in the industry, the time going forward is termed a "data decade." It is that time in the industry where technology advancement is happening at a pace beyond one's ability to stay in harmony with the evolution. However, what is interesting is the fact that key trends, which were perceived to be evolving in isolation, are seemingly coming together as perfect pieces of a puzzle.

Data engineering as a stream is getting sophisticated, and the cloud is making it real and relevant for enterprises. Working through the volume of data and establishing a story of insights and inferences is complex territory. To complement this construct, it is also an era where the bulk of the complexity involved in leading from data ingestion to inference is getting democratized. Isn't it such a great time to be a data engineer? The role of a data engineer is getting richer by day, and the need of the hour is to establish a certain acquaintance with cloud native offerings that take the power of insights and inference to an all new level.

This book is the coming together of Harsh Chawla and Pankaj Khattar, who bring in specialist knowledge in data and analytics. At Microsoft, they are consulted to solve key challenges of customers who are at a juncture to modernize their data infrastructure. The field of data and analytics is evolving with new platforms and techniques, and this book surfaces key concepts surrounding big data and advanced analytics. As I read through each chapter, it is a classic representation of the journey of data from inception to visualization, leveraging key services on Microsoft Azure. It was my privilege to offer content review support to Harsh and Pankaj, and their efforts are commendable, as they have made all efforts to present their perspective as practitioners.

FOREWORD

Authoring a book is no small feat, and kudos to Harsh and Pankaj for carefully crafting each chapter that brings out the nuances of the need and evolution of data lake and advanced analytics for modern businesses. It's a great read for any data engineer who wishes to upskill on technicalities behind setting up a data lake leading to advanced analytics on Microsoft Azure. It is also a great reference for data scientists to establish know-how of the overall data story on Azure. Congratulations to Harsh and Pankaj on this authorship and wishing them all the luck in taking their learning and sharing journey to greater heights.

Happy Reading...

Sandeep J Alur
Director, Microsoft Technology Center – India,
July 2020

# Acknowledgments

**Harsh Chawla** - I would like to thank my wife, Dharna, and my son, Saahir, for supporting me while I was writing this book. The book would not have been possible without you. I am indebted to our parents and families for motivating me to follow my dreams. I also am grateful to all of my mentors and the technical community for their guidance and inspiration. And I thank the entire Apress team for their patience and support. These acknowledgments would not be complete without me thanking Raj K, Gaurav Srivastava, Jerrin George, Sanket Sao, Venkatraghavan N and the entire EMEA DSD team for always being there for me.

ACKNOWLEDGMENTS

**Pankaj Khattar** - Writing a book is actually harder than I thought but more rewarding than I could have ever imagined.

None of this would have been possible without consistent support from my friend and wife Palak, from reading early drafts to giving me advice on the content and to managing our toddler daughter, so that I could concentrate and write this book.

I am eternally grateful toward my parents and family for imparting discipline, positivity, care, and passion, which have aided me in moving forward energetically. I am also thankful to my parent-in laws for their generous care, consistent support, and healthy advices.

I am also thankfull to my mentors, seniors, colleagues and friends. Specially to Pankaj Sachdeva, Dhiraj Bansal, and Gaurav Kumar Gupta for always encouraging me to be innovative and henceforth learn new things, which finally could go in this book.

A special acknowledgment & gratitude towards my senior leaders Dahnesh Dilkhush and Sujit Shetty for guiding, motivating and encouraging me, to belive and achieve above and beyond the limits.

A heartfelt thanks to my teacher Mr. RL Virmani, who taught me never to give up until I achieve what I really should.

A big thanks to the Apress team for their consistent support and direction toward finishing this book.

Finally, a delightful thanks to my lovely daughters Dakshita (a.k.a. Noni) and Navika (a.k.a. Shoni) for bringing unparallel joy and peace to my heart, soul, and life.

**We would like to thank Nidhi Sinha, Samarendra Panda, and Saurav Basu for their contribution to this book.**

# Introduction

Data is the buzz word today. Every organization is investing heavily to harness the power of data. Moreover, with the advancement of public cloud platforms accelerating the adoption of data analytics, the extremely complex scenarios like IOT, data lakes, and advanced data analytics have become uncomplicated and economical to manage. With these change, lots of learning and unlearning has also become really crucial. Henceforth, some new roles like data engineers and data scientists have shaped up. There is also a wider opportunity now available for database administrators or developers to upskill themselves into these roles.

There are typically two personas of database pros: relational DB pros and NoSQL DB pros. This book will be relevant for both these personas and others who are trying to get into the big data engineering field. The content of this book will help them to make their journey easier, as it has in-depth coverage and exercises for hands-on learning.

This book has been divided into two major sections for basic and advanced knowledge. In the first section, the discussion is on the types of data, evolution of enterprise data warehouses, and data analytics platforms. It also covers how modern data solutions and microservices applications have influenced the consolidation of these two solutions into one (i.e., modern data warehouse). In the second section, there is in-depth content on the multiple phases of data analytics solutions and respective technologies available in Microsoft Azure to build these scalable, distributed, and robust solutions.

This book also covers the infusion of AI/ML and real-time scenarios, and how they have evolved into advanced data analytics solutions. The architecture discussions in this book are cloud agnostic and can be utilized on different public cloud platforms.

We believe that this book will be a great resource for the technical community, and will help readers to learn more about big data engineering and advanced data analytics solutions.

# CHAPTER 1

# Data Lake Analytics Concepts

As they say, data is the new oil. This current decade has set a steep trajectory of data growth, and organizations have been greatly successful by harnessing the power of this data. Not only that, with the democratization of **A**rtificial **I**ntelligence and **M**achine Learning, building predictions has become easier. The infusion of AI/ML with data has given lots of advantages to plan future requirements or actions. Some of the classic use cases are customer churn analysis, fraud analytics, social media analytics, and predictive maintenance. In this chapter, the focus is on how the evolution of data happened and how it disrupted the data analytics space. Moreover, there will be a brief description of the concepts of relational and nonrelational data stores, enterprise data warehouse and big data systems.

## What's Data Analytics?

This is the first and foremost question that every data engineer or newcomers into the field of data analytics should know the answer to. Data analytics is a process of extracting, cleansing, transforming, and modeling data with an objective of finding useful information to help decision making. There are four major categories of data analytics:

1. *Descriptive*: What happened?

2. *Diagnostic*: Why did it happen?

3. *Predictive*: What will happen?

4. *Prescriptive*: What should I do?

© Harsh Chawla and Pankaj Khattar 2020
H. Chawla and P. Khattar, *Data Lake Analytics on Microsoft Azure*,
https://doi.org/10.1007/978-1-4842-6252-8_1

Traditional business intelligence (BI) systems have been catering to descriptive and diagnostic analytics, a.k.a. data analytics. However, when machine learning is added into the data analytics pipeline, predictive and prescriptive analytics becomes possible and it's called **advanced data analytics**, which is synonymous with traditional data mining solutions. Various use cases like fraud analytics, predictive maintenance, weather forecast, and optimizing production are examples of advanced data analytics. Data analytics has two major fields, which are called data Engineers and data scientists. To briefly describe these two fields, data engineering deals with extraction, transformation, and loading of data from data sources like relational and nonrelational. These data sources can be from IOT applications, ecommerce websites, or even finance applications. The data engineer needs to have a deep understanding of these data sources and the technologies that can help to extract and transform this data. However, data scientists use this data to build predictions using machine learning models and improve the accuracy of these models. Let's discuss relational and nonrelational data stores in the next section.

# Relational and Nonrelational Data Stores

Organizations have been building applications on both relational and nonrelational data stores. In fact, there are applications based on polyglot persistence–that is, multiple data stores like relational, key value pair, or document store in a single application. In other words, relational and nonrelational databases are two personas of data. The journey of relational will be completely different than nonrelational data; for example, to analyze large amount of relational data, enterprise data warehouses are built, or massively parallel processing is used. However, to analyze nonrelational data, data analytics solutions are built on distributed computing technologies like Hadoop. Moreover, now companies want to analyze both relational and nonrelational data on a single platform and build 360-degree analytics solutions. Let's explore what relational and nonrelational data stores are, as follows.

# Relational Data Stores

Relational data stores, a.k.a. (RDMS) **R**elational **D**atabase **M**anagement **S**ystems, have been built to manage structured data. These systems use ACID principle to run the database operations. The explanation of ACID principles is as follows:

> **Atomicity** means all the changes in a transaction will either be committed or rolled back.

> **Consistency** means that all the data in the database will be consistent all the time; none of the constraints will ever be violated.

> **Isolation** means that transaction data, which is not yet committed, can't be accessed outside of the transaction.

> **Durability** means that the committed data, once saved on the database, will be available even after failure or restart of the database server.

ACID principles are natively built into RDBMS database solutions, and it's much easier to manage the transaction processing with such solutions. However, scalability is the biggest challenge in RDBMS technologies. RDBMS systems are designed to scale up/scale vertically–that is, more computing can be added to the server rather than adding more servers. However, there are options to scale horizontally or scale out, but issues of write conflicts make it a less preferred option. Microsoft SQL Server, Oracle, PostgreSQL, and MySQL are examples of relational data stores.

# Nonrelational Data Stores

Nonrelational data stores contain semistructured and unstructured data. NoSQL data stores are types of semistructured data stores that are built on a BASE model. BASE is:

1. *Basically available*: The system appears to be available most of the time.

2. *Soft state*: The version of data may not be consistent all the time.

3. *Eventual consistency*: Writes across the services will be done over a period of time.

Technologies like IOT sensors and data streaming generate semistructured (JSON documents) data, and weblogs, videos, images, or clickstream data are all types of unstructured data that could be stored on Blob data stores. There are four different types of NoSQL databases, which are used for respective use cases:

1. *Document store*: MongoDB, Cosmos DB

2. *Column family store*: Cassandra, Cosmos DB

3. *Graph store*: Gremlin, Cosmos DB

4. *Key value pair*: Table, Cosmos DB

These data stores are distributed data stores, which follow the CAP theorem to run database operations. CAP theorem has three components (Figure 1-1):

> *Consistency*: Every read must receive the most recent write or an error.

> *Availability*: Every request must receive a response, without a guarantee of it being the most recent write.

> *Partition tolerance*: The system must continue to work during network failures between components–that is, the system will continue to function even if one of the nodes fails.

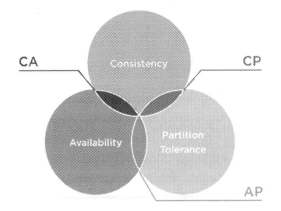

***Figure 1-1.*** *CAP theorem*

CAP theorem describes that any distributed application can achieve only two functionalities out of these three, at the same time. Depending upon the nature of the application, one can choose an intersection of two functionalities: consistency and partition tolerance, or consistency and availability, or availability and partition tolerance. Most NoSQL technologies implement eventual consistency and prefer availability over consistency. However, this is customizable, and one can choose any two functionalities depending on the nature of application. Therefore, a NoSQL database can support more than one thousand nodes in a cluster, which is not possible with RDBMS databases.

# Evolution of Data Analytics Systems

A decade ago, the majority of applications were monolithic. These solutions were largely built on top of structured data stores like SQL server and Oracle, etc. Further to that, NoSQL technologies created buzz in the software market. Moreover, there was an evolution in application architectures from monolithic to SOA (service-oriented architecture) and now microservices, which promoted the use of polyglot technologies. That's when the tipping point started; data of multiple types was getting generated from a single application. Moreover, there were multiple independent LOB (line-of-business) applications running in silos. Similarly, CRM (customer relationship management) and ERP (enterprise resource planning) applications generated lots of data, but systems were not talking to each other. In other words, not only the variety of data but the velocity and volume of data were significantly large. There were two major methods to analyze large amounts of data:

1. Enterprise data warehouse (EDW)

2. Big data analytics

# Enterprise Data Warehouse

An enterprise data warehouse has been known as a solution to analyze large amounts of structured data. This solution has been working for decades. As shown in Figure 1-2, in conventional data warehousing systems, standard procedure to analyze large data was as follows:

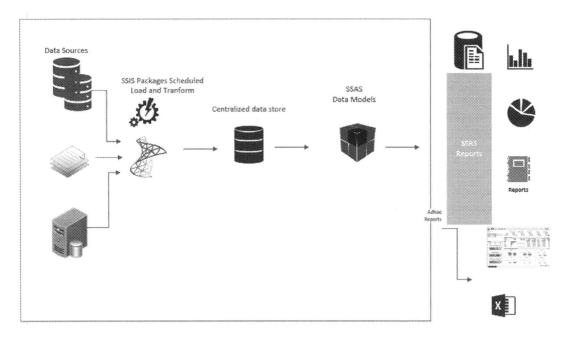

***Figure 1-2.*** *EDW architecture built on SQL server*

1. Consolidate data from different data sources using ETL tools like SSIS (SQL Server Integration Services) packages.

2. Build centralized data stores to feed data to SSAS (SQL Server Analysis Services) data models.

3. Understand KPIs and build data models.

4. Build a caching layer using OLAP(SSAS) platforms to precalculate KPIs.

5. Consume these KPIs to build dashboards.

However, due to SOA or microservices applications and especially with new types of data like JSON data, key value pair, columnar, and graph, there was a need that standard enterprise data warehouses were not able to fulfill. That's where big data technologies played a key role.

# Big Data Systems

Big data systems were specialized to solve the challenge of the four Vs: volume, variety, veracity, and velocity. The scenarios like gaming, social media or IOT there incoming data streams can be from millions of users and can be fast and of large volume. Let's examine the four Vs in a little more detail as follows:

### Volume

Platforms like gaming, ecommerce, social media, and IOT, etc. generate large amounts of data. Let's take an example of IOT: connected cars where a car manufacturing company may have millions of cars across the globe. When these cars are sending telemetry to the back-end systems, it can be of high volume; it can be hundreds of GBs of data per day. That's the volume challenge that big data systems solve.

### Variety

Traditionally, data has been in a structured format of data types like text, date, time, or number. To store this type of data, relational systems have always been a perfect fit. However, with the emergence of unstructured data like Twitter feeds, images, streaming audios, videos, web pages, and logs, etc., it was challenging to have a data analytics system that can analyze all these types of data. That's the variety challenge that big data systems solve.

### Veracity

Data veracity is another critical aspect, as it considers accuracy, precision, and trustworthiness of data. Given that this data would be used to make analytical decisions, it's something that should always be truthful and reliable. Veracity of data is really critical to consider before designing any big data systems. Since the entire solution is based on data, understanding validity, volatility, and reliability of data is critical for meaningful output.

**Velocity**

Data velocity refers to the high speed of incoming data that needs to be processed. As discussed earlier, with scenarios like IOT, social media, etc., speed of incoming data will be high and there should be systems available that can handle this kind of data. That's the velocity challenge that big data solves.

To support the aforementioned challenges of the four Vs and cost effectiveness, big data technology platforms use distributed processing to build scalability, fault tolerance, and high throughputs. In distributed processing, there are multiple different compute resources that work together to execute a task. Hadoop is a big data solution that is built on distributed processing concepts, where thousands of computer nodes can work together on PBs of data to execute a query task. For example, HDInsight (managed Hadoop on Microsoft Azure) and Databricks are heavily used platforms on Microsoft Azure.

So far, there has been a brief mention of enterprise data warehouse and big data concepts. While big data technologies were emerging, there was another concept called massively parallel processing (MPP) that came into light. It helped to analyze large amount of structured data. However, building an enterprise data warehouse took a long time considering the ETL, building centralized data stores, and OLAP layer, etc. involved. This solution was cumbersome, especially when the data size was in dozens to hundreds of TBs. There was a need for technology that could analyze this amount of data without the overhead of going through the long cycle. With MPP technologies, SQL queries can be executed on TBs of data within seconds to minutes, which could take days to run on traditional EDW systems.

# Massively Parallel Processing

In MPP systems, all the compute nodes participate in coordinated operations, yet are built on shared nothing architecture. These nodes are independent units of compute, storage, and memory; to reduce the contention between nodes, there is no resource sharing between the nodes. This approach relies on parallelism to execute the queries on large data. In MPP, all the data is split into multiple nodes to run the queries locally and in parallel with all the other nodes. This helps to execute queries faster on TBs of data. Examples of MPP solutions are SQL pools in Synapse Analytics (earlier known as SQL DW) and Redshift, etc.

So far, there has been a discussion of three major types of analytics systems: enterprise data warehouse, big data, and MPP systems. Now, let's understand where the word data lake analytics comes from.

# Data Lake Analytics Rationale

Within the aforementioned three choices, organizations chose the most suitable one and built their solutions. With the advent of new applications and demand to build 360-degree analytics scenarios, there was a need to consolidate all the data under one platform. Especially, data scientists wanted a single pool of data that they could use to build predictive and prescriptive solutions. It's difficult for organizations to manage an enterprise data warehouse for structured data and manage big data systems for the rest of the unstructured and semistructured data. It's like breaking silos and creating another silo within analytics solutions. The concept of bringing all the data–structured, semistructured, and unstructured–and building a central store, is called a data lake. Building analytics on top of this data is called data lake analytics, and it's also called data analytics. This concept has helped to merge enterprise data warehouse and big data analytics solutions into a single solution called a modern data warehouse. Moreover, when AI/ML is infused into a modern data warehouse, it becomes advanced data analytics. Figure 1-3 shows both modern data warehouse and advanced analytics merged into one.

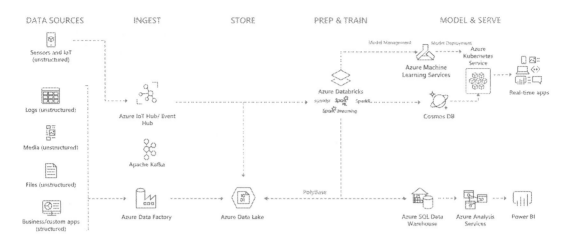

***Figure 1-3.*** *Modern data warehouse and advanced analytics architecture*

In modern data analytics solutions, the entire solution is divided into five major sections:

1. Data sources

2. Data ingestion

3. Data storage

4. Preparation and training

5. Model and serve

In the coming chapters, these steps are discussed in detail.

# Conclusion

This chapter is about the basics of data analytics and how the evolution of data analytics platforms happened. The chapter is important for both data engineers and other DB profiles who are trying to get into this field. It briefly covers relational and nonrelational data stores and how the application landscape has changed from monolithic to microservices architectures. Moreover, there is a discussion on how these changes have brought a shift from enterprise data warehouses to big data analytics and massively parallel processing solutions. Fast-forward to the cloud world, where the enterprise data warehouse has been merged with big data analytics solutions into modern data warehouses; and when ML is infused into this solution, it becomes advanced analytics solutions. In the next chapter, there is a deeper discussion of the basic building blocks of data analytics solutions.

# CHAPTER 2

# Building Blocks of Data Analytics

Data analytics solutions have become one of the most critical investments for organizations today. Organizations that are ahead of the curve have been spending lots of money, time, and resources on this practice. Major tech giants like ecommerce, social networking, and FMCG (fast-moving consumer goods) are heavily dependent on their data analytics solutions. A few examples of the outcomes of data analytics are customer 360-degree, real-time recommendations, fraud analytics, and predictive maintenance solutions. This chapter is designed to share an overview of the building blocks of data analytics solutions, based on our learning from large-scale enterprise projects and helping organizations to start this practice.

## Building Blocks of Data Analytics Solutions

Data analytics is a process of extracting, cleansing, transforming, and modeling data with an objective of finding useful information to help in decision making. Data analytics is done in a series of steps to extract value out of data. There are multiple components that come together to build this engine. Let's discuss some of the critical components as shown in Figure 2-1:

1. Data sources

2. Data ingestion

3. Data storage

4. Preparation and training

5. Model and serve

© Harsh Chawla and Pankaj Khattar 2020
H. Chawla and P. Khattar, *Data Lake Analytics on Microsoft Azure*,
https://doi.org/10.1007/978-1-4842-6252-8_2

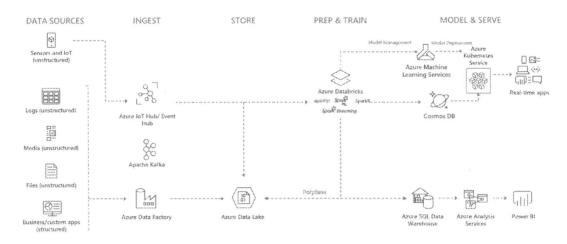

***Figure 2-1.*** *Building blocks of a data analytics engine*

# Data Sources

A data source is a storage system where the input data resides. As highlighted in Figure 2-2, there are three major types of data sources, and all the products or applications fall under these categories.

1. Sensors and streaming data

2. Custom applications

3. Business applications

**Sensors and streaming data** - Telemetry data coming from IOT, data streams coming from Twitter, social media websites, clickstream data, etc. is generated in real time and falls under this category.

**Custom applications –** Data coming from custom applications is built on relational and nonrelational data stores. Moreover, the media files like videos and pictures, log files, etc. belonging to an application are part of this category.

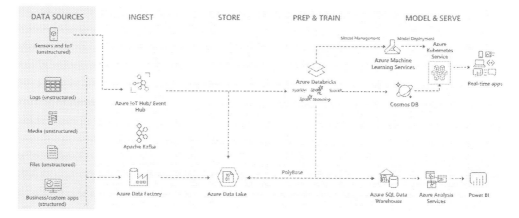

***Figure 2-2.***  *Data sources*

**Business Applications –** SaaS applications Dynamics 365, Google Analytics, Adobe, and Inmobi marketing applications fall under this category.

To fetch data from these data sources, there are various technologies available under the data ingestion section of data analytics solution architecture. Let's discuss these technologies briefly in the next section.

# Data Ingestion

Data ingestion is an important and critical step in a modern data warehouse or advanced analytics system. This is the first step in building the data analytics solution. As highlighted in Figure 2-3, there are various options available to get the data ingestion done.

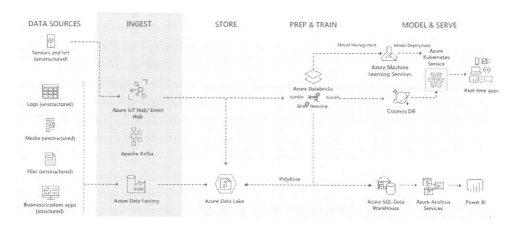

***Figure 2-3.***  *Data ingestion*

There are two categories of data ingestion technologies:

1.  Real-time data ingestion

2.  Batch model data ingestion

**Real-time data ingestion** – Data coming from the IOT and streaming applications fall under this category. There are two architecture patterns to ingest real-time data:

1.  Lambda architecture

2.  Kappa architecture

**Lambda architecture** – Lambda architecture defines the way high-velocity incoming data should be processed. Let's take an example of an IOT scenario where temperature sensors from an oil exploration plant are sending data in real time. In case of an emergency, the alarm should be raised in real time. Raising the alarm after the tragedy happens defeats the purpose. Lambda architecture has defined two data paths, as referred to in Figure 2-4:

1.  *Hot data path*: Speed layer

2.  *Cold data path*: Batch layer

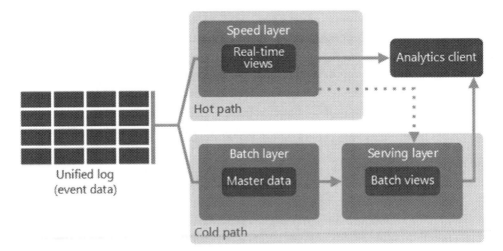

**Figure 2-4.** *Lambda architecture*

**Speed Layer** - Streaming data ingestion is essential for real-time data processing. The data is absorbed by the ingestion layer as soon as it is available at the source. Either the source can push the data or it could be pulled by the ingestion layer into the data pipelines. Think of it as continuous flow of data coming from a data pipeline. This type of data ingestion strategy is typically very expensive, as it requires consistent monitoring of source systems and the ingestion layer must be scalable enough to absorb dynamic flow of data to avoid throttling of pipeline capacity.

**Batch Layer** - Batch data ingestion for batch processing is the most common practice. In this layer, data is collected from the data sources at regular intervals and pushed to the destination storage. The intervals can be scheduled, triggered on demand, or on execution of any external event. Batch data ingestion helps in performing batch and near real-time analytics. It is the easiest, least time consuming, and most cost-effective data ingestion strategy.

**Kappa Architecture** - There is another architecture pattern to ingest real-time data. This architecture removes the batch layer completely. This architecture was proposed by the LinkedIn team that created Apache Kafka. Kappa architecture supports a composite data ingestion strategy called micro batching data ingestion. It is a combination of both batch and streaming strategies, where streaming data is grouped in multiple continuous batches for ingestion.

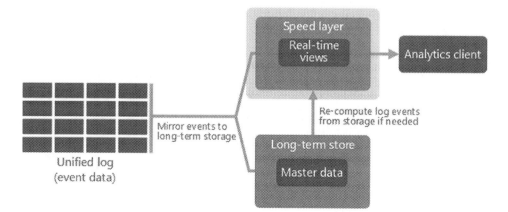

***Figure 2-5.*** *Kappa Architecture*

There are multiple technologies available to ingest data from various data sources; however, there are two strategic challenges that should be considered while designing the solution.

> **Challenge 1** - As the data grows at a rapid pace, data sources also evolve at a rapid pace. These data sources are transaction databases, IOT devices, mobile and web applications, etc., which continue to evolve, grow, and become more sophisticated with new features. Therefore, defining a stable and consistent data integration strategy is extremely difficult to implement. Maintaining an analytical platform that performs the ingestion of a high volume, high velocity, and high variety of data is complex, costly, and time consuming.

> **Challenge 2** – As the data volume grows, building data pipelines becomes more complex and tend to go slow when it reaches throttling capacity. Hence, pipelines need to be scalable and elastic so that they adapt to change in velocity and volume of data for real-time data ingestion.

Generally, at the start of a project the organizations are not sure if real-time data processing is needed. Therefore, to save cost they simply adopt a batch processing data ingestion strategy. However, if there is need to scale up for real-time data processing, they must make drastic architectural changes. So, it's important to choose the right set of technology that is agile, elastic, and scalable.

As shown in Figure 2-6, some of the examples of data ingestion technologies are as follows:

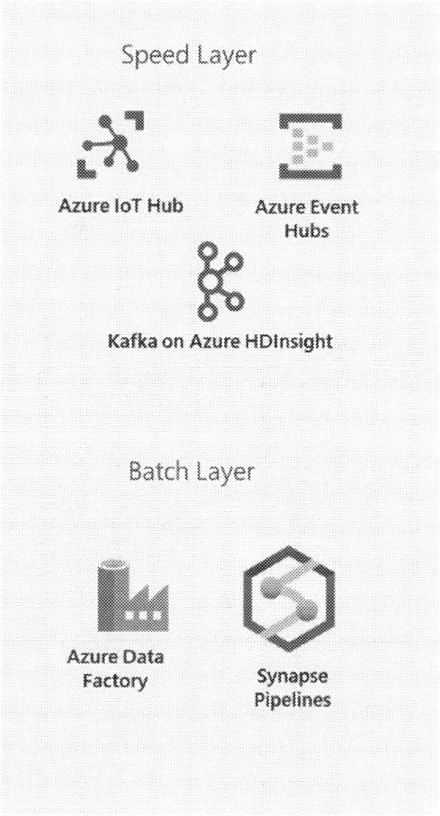

**Figure 2-6.** *Data ingestion stack*

Speed layer, a.k.a. real-time data ingestion:

1. Azure IOT hubs

2. Event hubs

3. Apache Kafka

Batch layer a.k.a. batch mode data ingestion:

1. Azure Data Factory

2. Azure Synapse pipelines

# Data Storage

The data storage layer lays the foundation for building a robust and scalable enterprise data lake. It should be designed from the beginning to service a massive amount of data and provide high-end consistent throughput without manageability overheads.

The storage generally should have the following features:

- *Accessible*: Data stored in storage devices should be accessible from any location on a public/private network. It should ideally support the open protocols like HTTP/HTTPS and provide an SDK to integrate the services in the applications consuming data.

- *Scalable*: Given the data volume continues to scale, it's vital that the storage device is scalable and provides an optimum level of performance for multiple reads and writes.

- *Highly available*: In case of any machine, hardware, or data center failure, there should be a capability available to replicate the data across multiple data centers in different geographical regions. In case of any power failure or catastrophe, data is still available for consumption.

- *Fully managed*: Fully managed storage can save lots of maintenance efforts and cost, as the hardware maintenance, software upgrades, debugging issues, and support activities are done by the storage service provider.

- *Secure*: Data encryption is important for data safety. Encryption should be both for data at rest and in motion. Encryption should also be supported by role-based access controls so the user can only view or edit data on which they have authorization.

- *Hierarchical namespaces*: A hierarchal namespace allows objects and files to be stored in a directory or subdirectory like the filesystem of a computer. With this feature, the storage becomes capable of providing scalability of an object storage and semantics required for analytics platforms.

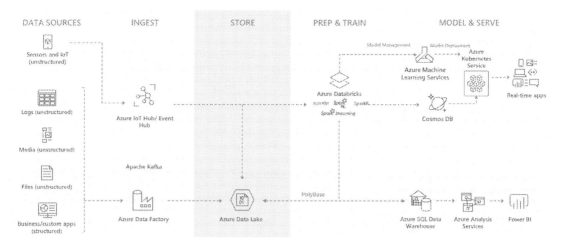

***Figure 2-7.*** *Storage options*

As shown in Figure 2-7, there are two options available on Microsoft Azure platforms.

1.  Azure Blob Storage

2.  Azure Data Lake Storage

**Azure Blob Storage** - Azure Blob Storage is Microsoft's object storage solution for the cloud. Blob storage is optimized for storing massive amounts of unstructured data. Apart from storing huge amounts of data, it can serve pages to a browser directly, provide distributed access to the files, stream video/audio, and archive data. This has always been a preferred storage solution for various data stores and data analytics solutions. However, there was a need to build something specific for data analytics, where the data needs to be accessed faster with GBs of throughput and can scale to PBs. On Microsoft Azure, Azure Data Lake Storage is dedicated to such workloads and it's built on HDFS principles.

**Azure Data Lake Storage** - Azure Data Lake is a cloud-based storage layer designed for big data analytics in Microsoft Azure cloud. It provides features like hierarchical file access, directory- and file-level security, and scalability combined with low cost, tiered storage, high availability, and disaster recovery capabilities

Data lake storage enhances performance, management, and security as follows:

*   Performance is optimized, as no data transformation is required prior to the analysis of data. Hierarchical access greatly improves file and directory access and management, which results in overall improvement in job execution.

- Data management becomes relatively easy as the data can be organized in files and folders.

- Data security is defined using POSIX (Portable Operating System Interface) permissions on directories and files.

# Data Preparation and Training

In a traditional enterprise data warehouse, data transformation was the preferred way of altering structure, formats, and multiple values of the incoming data. However, in modern data analytics solutions–where the data is coming from disparate data sources, in different formats, and in real time or in batches–data transformation alone doesn't meet the purpose. Therefore, data preparation and training has been a new term introduced in modern data warehouses and advanced data lake analytics solutions. As shown in Figure 2-8, the discussion is on data preparation and training in this section.

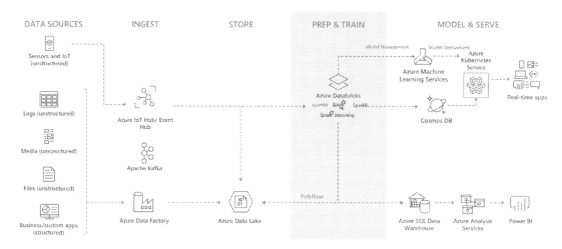

***Figure 2-8.***  *Data preparation and training*

Moreover, with the cloud being the preferred platform for analytics solutions, there has been a change in the term ETL (extract, transform, and load) to ELT (extract, load, and transform) as well. First, let's get into the details of ETL- and ELT-based solutions (Figure 2-9); it will help to relate better with the new term "data preparation and training."

**ETL** - In the ETL-based solutions, the data is extracted or read from the multiple data sources as the first step. Then the read data is transformed from one format to another format. The transformation is usually more of a rule-based transformation or joining of data from multiple sources by performing lookups from external sources. Once transformation is performed, data is loaded into a target database or data warehouse.

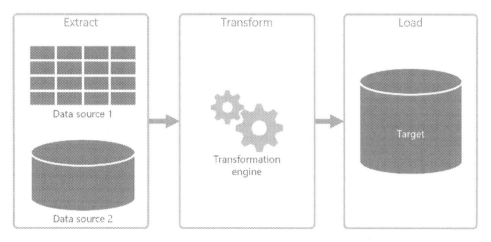

***Figure 2-9.*** *ETL process*

**ELT** – ELT-based solutions have become very popular after the public cloud came to light. Since the variety, velocity, and volume of data are different in modern data analytics solutions, it's impractical to transform the data while ingesting. It can cause throttling and bottlenecks at the source systems, especially when the data is being pushed into the ingestion pipelines. Since storage is inexpensive and a data warehouse can be invoked on demand, the transformation can be run for few hours with large compute resources. It serves two purposes:

1.  Since the public cloud offerings are pay as you go, the charges will be applicable only for the duration of the job run.

2.  To increase the speed of processing, compute resources can be scaled up.

Therefore, to save cost and have better performance, data is directly loaded into either cloud storage or a data warehouse first and then transformed. Data transformation is a subset of data preparation.

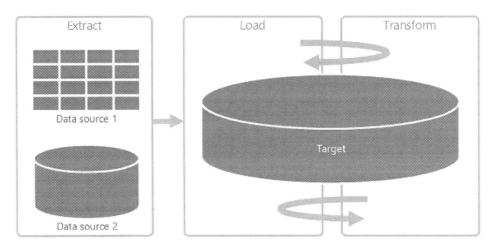

Data preparation is the process to improve data quality, to make the raw data ready to be analyzed—that is, use it to train ML models or send it directly to the model and serve phase. It consists of following steps:

1. *Data cleaning*: This is a process to detect and remove noisy, inaccurate, and unwanted data.

2. *Data transformation*: This normalizes data to avoid too many joins and improve consistency and completeness of data.

3. *Data reduction*: Data reduction is aggregation of data and identifying data samples needed to train the ML models.

4. *Data discretization*: This is a process to convert data into right-sized partitions or internals to bring uniformity.

5. *Text cleaning*: This a process to identify the data based on the target use case and further cleaning the data that is not needed.

As shown in Figure 2-8, data in the preparation and training phase comes in the form of real-time data streams or in batch mode. Based on these two categories, the technology solution to process the data will change. However, training deals with building, training, and improving the accuracy of an ML model. Therefore, it's preferred to be done only in batch mode.

For real-time data processing, the technology options are:

1. Apache Spark Streaming

2. Azure Stream Analytics

For batch mode data processing, the technology options are:

1.   Apache Spark

2.   Azure HDInsight clusters

3.   Azure Databricks

After the data is prepared and trained, it's sent to the model and serve phase.

# Model and Serve

The data model and serve layer is the abstraction between the data visualization layer and the output of massive work done behind the scenes to bring sense out of data. This is the destination of data after preparation and training is performed. Based on how the output data will be consumed and the downstream applications, technology on this layer is decided. In this section, let's examine the model and serve phase as highlighted in Figure 2-10.

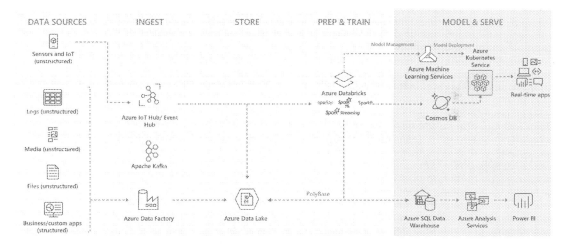

***Figure 2-10.***  *Model and serve phase*

Let's take a few examples to explain it better:

1.   *Enterprise data warehouse*: Let's take an example of structured data that gets stored in relational stores. For relational data, the data will preferably land in a data warehouse or an MPP solution. As shown in Figure 2-11, Azure Analysis Services is preferred for caching the data models.

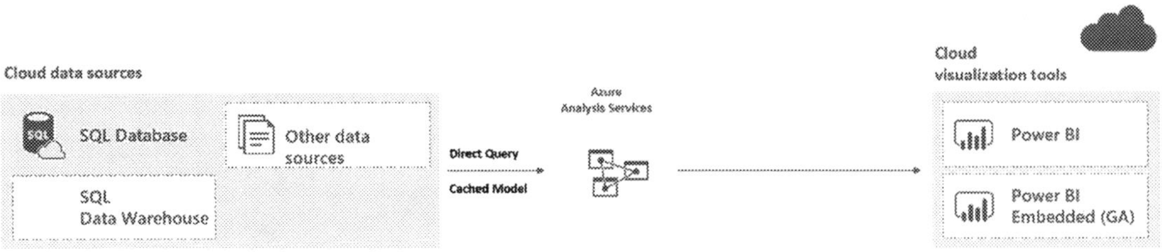

**Cloud data sources**

SQL Database

Other data sources

SQL Data Warehouse

Direct Query

Cached Model

Azure Analysis Services

**Cloud visualization tools**

Power BI

Power BI Embedded (GA)

***Figure 2-11.*** *Data serving layer for relational stores*

2. *Data lake analytics*: In data lake scenarios, the data can come from the hot path as well as the cold path. If the data is coming from the hot path, there is a high possibility that data would be needed to serve in real time. In this scenario, this data can be directly used to build dashboards in Power BI or use Cosmos DB change feeds to raise alerts using Azure functions, etc. However, if the data is coming from the cold path where the data will be processed in batch mode, Synapse Analytics SQL pools for data processing, Azure analysis services for caching KPIs, and Power BI for dashboards are preferred options.

# Data consumption and data visualization

This layer serves the purpose of the entire data analytics pipeline. This layer is consumed by businesses to understand the customer 360, fraud analytics, predictive maintenance, and other downstream business applications and dashboards.

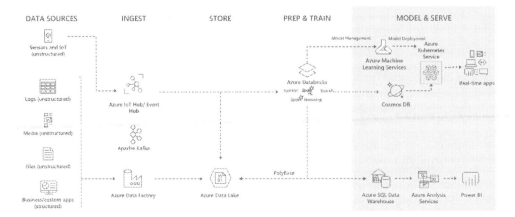

***Figure 2-12.*** *Data Visualization*

Let's talk about a few scenarios to understand it better.

1.  *Customer churn analysis*: Let's take an example of customer churn analysis. Telecom providers run this analysis to understand which customers are most likely to churn; depending on the tier of the customer, various offers are rolled out to the marketing team. This helps the companies with customer retention. The input data to the marketing application comes from this pipeline. Azure SQL DB can be a source of information to the marketing application. Similarly, the same information can be provided to the call center employees on their CRM applications to ensure better customer satisfaction.

2.  *Stock market prediction*: There are scenarios where stock brokers can run algorithms on an upcoming stock that will yield high profit, and notify their end customer through their application. This can be a notification service of the application, which consumes data from Azure Cosmos DB.

3.  *Dashboards*: This is one of the most common scenarios, where the business wants to know how they are progressing and upcoming strategies of expansion or more KPIs need to be projected on a dashboard. They can fetch data from SQL DW or DB or Cosmos DB, etc. Power BI, Tableau, and QlikView are the solutions for dashboards

# Conclusion

In this chapter, the focus was to share how the entire lifecycle of data is managed in a data analytics solution. All the major concepts that are important for a data platform architect or data professional have been discussed in this chapter. The next chapter delves deeper into each phase of this lifecycle and how Microsoft Azure really helps to optimize cost yet build efficient data analytics solutions.

# CHAPTER 3

# Data Analytics on Public Cloud

Public cloud has been disrupting businesses in a big way. Data analytics is another area where a public cloud has brought lots of innovation. Organizations can build, scale, and consume these solutions with a faster pace and economical cost. Since there is a fair understanding of data lakes and data analytics basics by now, this chapter discusses the role of a public cloud to disrupt the market and to accelerate the adoption of data analytics solutions.

## Reference Architectures on Public Cloud

Public cloud adoption has grown multifold in the last couple of years. Organizations have realized the potential of cloud computing. Scalability, agility, security, and choice of technology platform on a public cloud are commendable. Microsoft, Amazon, and Google are the major cloud service providers in the market. There are three ways to host a data analytics solution:

1. Traditional on-premises

2. Infrastructure as a service

3. Platform as a service

© Harsh Chawla and Pankaj Khattar 2020
H. Chawla and P. Khattar, *Data Lake Analytics on Microsoft Azure*,
https://doi.org/10.1007/978-1-4842-6252-8_3

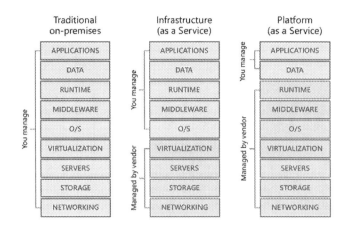

***Figure 3-1.*** *Deployment models from on-premises to public cloud*

# Traditional on-premises

This has been a conventional way to host data analytics or any IT solution, so far. It's the private data center where an entire data analytics pipeline is set up on VMs or physical machines. It takes weeks and huge upfront investment (capital expense) to set up the solution. Even to set up the big data clusters, installation and configuration can take long time. Therefore, it becomes the blocker for many organizations. However, a public cloud like Azure could offer many benefits in this situation. The majority of the components are very easy to set up, and within a few minutes the entire setup can be completed. There is no need for upfront investment; it's per minute/hour billing where companies can pay as per the usage of resources (operational expense).

Let's discuss two major ways to build resources on a public cloud.

# Infrastructure as a service (IaaS)

IaaS is typically managed in the same way as your own private datacenter. As per Figure 3-1, the end user needs to manage everything above the virtualization layer. Virtual machines are backed by uptime SLAs. Based on the criticality of the application and acceptable downtime, VMs need to be configured.

| Deployment Type | SLA | Downtime per Week | Downtime per Month | Downtime per Year |
|---|---|---|---|---|
| Single VM | 99.90% | 10.1 minutes | 43.2 minutes | 8.76 hours |
| VM in Availability Set | 99.95% | 5 minutes | 21.6 minutes | 4.38 hours |
| VM in Availability Zone | 99.99% | 1.01 minutes | 4.32 minutes | 52.56 minutes |

For example, for the high availability of the SQL server instance, the virtual machines should be in the availability set. The availability set is the logical grouping of virtual machines allocated in an Azure data center. The availability set ensures that the VMs are set up in the same data center but on different fault and update domains. The fault domain ensures the resources in availability sets are created on multiple physical servers, compute racks, storage units, and network switches. The update domain ensures the nodes in availability sets are updated separately. Moreover, the availability zone is another offering that protects the applications from the datacenter failures. Depending on the uptime need, a suitable option could be selected.

## Platform as a service (PaaS)

With PaaS, the entire infrastructure is managed by the cloud service providers. As described in Figure 3-1, administrators just need to manage their applications and data; the rest—everything like hardware, virtualization, and runtime software—is managed by the cloud service providers. Therefore, managing resources becomes convenient for administrators. Let's take an example of Azure SQL DB; the entire OS layer and SQL server layer is managed by Microsoft. Backup, monitoring, and security are automatically done at a basic to moderate level, and customization is available at the click of a button. Database experts just need to manage the databases, respective application schemas, and code, etc.

Let's discuss the various challenges faced by decision makers for IT investments and how a public cloud solves those challenges:

1. Cost

2. Scalability

3. Agility

4. Manageability

5. Security

    1. *Cost*: This is one of the biggest decision influencing factors. No matter if the investment is on a public cloud or private cloud, the key spend is on:

        a. Hardware infrastructure

        b. Software licenses

        c. Security

        d. Product support of respective products

        e. People

        f. Technical readiness

    Setting up everything on a private cloud means upfront cost, which means longer approval cycle, longer setup time, and longer start time to execute. Moreover, we have seen that by the time planning and approval of a project is done, a new version or new or sometimes cheaper technologies arrive on the market. More than 50% of the time, by the time a project completes, the solution is already running on older technologies.

    Apart from this, buying hardware, software licenses, security, product support for the software, and then people and their readiness are the most complex tasks leadership need to solve. When a public cloud like Microsoft Azure is chosen instead of a private cloud, more than 60% of the hassle is taken care of due to the following reasons:

        1. Projects can be started with minimal investment, as the billing is done as per usage. However, there are options to save more cost by reserving the resources.

2.  There are options available to reduce dependency on software licenses; for example, data analytics technologies are mostly based on PaaS where per minute/hour billing is done and there is no need to buy separate licenses. Even if they are needed, monthly subscription-based licenses can be purchased. Moreover, if the licenses are already available, a BYOL (bring your own licenses) option can be used. This way already procured licenses can be used, which can further reduce the cost on Microsoft Azure.

3.  Unlike a private cloud, where support for each product must be bought separately, with Microsoft Azure only one support contract for the majority of products is sufficient.

4.  Security on a public cloud is a joint responsibility. However, public cloud infrastructure provides a certain level of security and capabilities natively in the form of services like DDoS and firewalls, etc. Moreover, there is a seamless integration with other options available from a variety of different vendors, which makes a public cloud a great option.

5.  Most of the data analytics products on Microsoft Azure like HDInsight, Azure Databricks, Azure Synapse Analytics, etc. provide multiprogramming support (i.e., C#, Java, Node.JS, etc. programming languages can be used). This solves the challenge of technical manpower, since people from various technical backgrounds in existing teams can be leveraged with minimal training.

2.  *Scalability*: Scalability is one of the major concerns on private data centers. It can be resolved with the help of virtualization technologies to an extent; however, it needs lots of technical expertise and time to make it happen. Some of the biggest challenges while sizing any application are predicting the scale of the application and having an experienced architect to get the hardware sizing right. Most of the time, applications are either undersized or oversized, which makes organizations either spend more or face performance issues during ad hoc/peak load. This challenge is solved by a public cloud seamlessly.

31

Public cloud platforms provide this functionality as a key feature. This helps to start the project with minimal infrastructure and scale up/down depending on the load on the application. In the context of a data analytics solution, auto-scalability is one of the major advantages of a public cloud. For data transformation that happens during the night or any specific interval anytime during the day, infrastructure can be allocated for a specific interval and can be scaled up to make the jobs run faster. After the transformation is done, the resources can be released completely. Moreover, this functionality helps to save lots of cost and gives more certainty to run the solution.

3.  *Agility*: The other challenge that companies face is agility in the solution. How quickly a technology product is replaced or added in a solution can be a big differentiator. For many technologies, software licenses may need to be procured, which again would need a round of approvals and upfront investment. This capability decides the rate of innovation in an organization. Today, if there is a great idea, organizations want to try a pilot very quickly, understand the potential, and decide the next action. Private data centers can't provide this sort of agility even with huge IT investment. The latest hardware like GPUs/FPGA/TPUs needs lots of upfront investment; even setting up ML platforms is a big hassle.

On Microsoft Azure, there is a range of technology solutions across the spectrum that can be used as pay as you go models. Months of setup time can be reduced to days or even hours. This enables the organizations to quickly innovate and build prototypes. Not only that, even for projects that are long term and fully operational, this capability helps to add new functionalities quickly. On Microsoft Azure, there are hundreds of new features and products added every month, which brings lots of flexibility for the end user.

4. *Manageability*: Manageability is another overhead for organizations. Initial set up time, software upgrade, and applying security fixes take a large number of man-hours investment for any organization. Considering the security vulnerabilities these days, new security fixes should be applied regularly or even software upgraded many times. These are additional operational overheads if the applications are hosted on a private cloud. Solutions like data analytics need to continuously evolve, and it becomes difficult to manage the solutions on-premises.

   With a public cloud, handling agility becomes comparatively easy. Most of the data analytics solutions on Microsoft Azure are PaaS based, which means an administrator doesn't need to worry about patches or upgrades. Everything is managed by Microsoft. Moreover, for advanced analytics—especially when dealing with video and image data—there can be requirement of GPUs and FPGAs processors, which are available for use in Microsoft Azure. Had this been an on-premises setup, this project would have needed lots of investment and taken lots of time to start the project.

5. *Security*: Security is the top priority for companies today. Since there are lots of incidents of data theft, hacking, or phishing, it's important to have security up to the mark. Companies are hiring CISOs (chief information security officers) today to ensure proper security standards are followed at each level. There are six major areas where companies spend their money:

   1. Identity and access management

   2. Network

   3. Application and data security

   4. Threat protection

   5. Intelligent edge (IOT devices)

   6. Security management

Setting up the security is a complex task, and upkeep is even more complex on private on-premises data centers. On a public cloud, most of these products are available on a pay as you go model. Moreover, public cloud companies have experience managing the applications at a global level; they also ensure innovation in the security products. The customer gets the choice of the latest security products available on the cloud platform.

By now, readers have a fair understanding of the data analytics ecosystem and why a public cloud is a great choice to build these solutions. Let's discuss the various architecture choices architects could make while building data analytics solutions. These are standard cloud design patterns for such solutions and can be applicable to any public cloud provider by just changing the product name. There are many open source products like Apache Spark, Apache Kafka, etc. that are available on any public cloud platform. Since this book is about building these solutions on Microsoft Azure, we will share the solutions available on the Microsoft Azure platform.

# Data Analytics on Microsoft Azure

The purpose of this chapter is to highlight the advantages of building a data analytics solution on a public cloud, especially Microsoft Azure. In the previous chapter, the discussion was around the building blocks of a data analytics solution, as mentioned in Figure 3-2, which is a generic architecture.

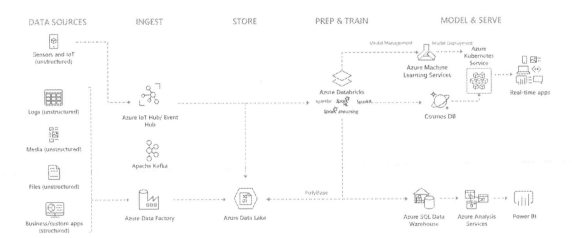

***Figure 3-2.***  *Data analytics pipeline*

Since, we are shifting gears to building the solution on Microsoft Azure, there will be a slight change to the aforementioned architecture. All the components will remain the same, but the architecture will be refined to make it easier to understand. Moreover, there will be a discussion around different architecture design patterns, which will help to choose right architecture based on the different scenarios. As mentioned in earlier chapters, there are two major design patterns that are generally used:

1.  Modern data warehouse

2.  Advanced data analytics

Let's go deeper into these architecture patterns and what scenarios each of these fit in.

### Modern data warehouse

Data is growing at a much larger scale, faster pace, and with multiple data types. It's important to bring all the data together under a common platform and fetch insights through dashboards. As shown in Figure 3-3, a modern data warehouse is a consolidation of enterprise data warehouse and big data analytics platforms. On top of that, it provides a unified platform to all the users for their analytical needs.

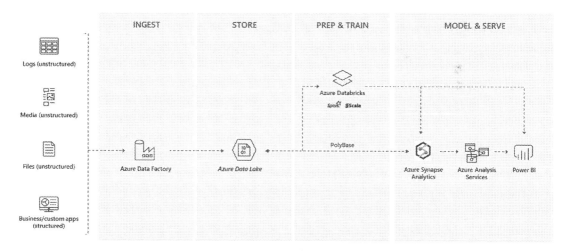

**Figure 3-3.**  *Modern data warehouse architecture*

Let's discuss a couple of advantages of building a modern data warehouse:

1. It helps in storing all your data–which could be structured, semistructured, or unstructured–at a centralized location. This centralized location is elastically scalable to save data, which can range from a few GBs to TBs of data without affecting the overall performance for reading and writing the data. Azure Data Lake Storage and Azure Blob Storage are such examples.

2. Once the data is available in the centralized location in a data lake, data preparation and transformation can be easily performed on that data using big data technologies like Hadoop or Spark. These engines can easily process the data in a distributed, scalable, and fault tolerant way. Azure provides components like Azure HDInsight and Azure Databricks, which are designed to perform the analysis.

3. Once the data is transformed and cleansed, it can then be moved to a data warehouse solution for OLAP requirements, where it can be combined with all the existing data for creation of a data hub. Azure provides Azure Synapse Analytics, which is a limitless analytical service that can deliver insights from all the data across data warehouse and big data systems with a very high degree of performance.

4. Operational reports and analytical dashboards can be built on Azure Synapse to further derive insights and present that to the business stakeholders. Azure Power BI can be easily integrated with Azure Synapse and Azure Analysis Services.

5. A business intelligence tool like Power BI can also execute ad hoc queries directly on an analytical platform like Databricks to fetch any real-time data or raw data.

Customers who have invested in building modern data warehouse platforms have seen the following benefits:

1. Data analytical reports are now available in minutes instead of days.

2. IT productivity is enhanced by utilizing the latest tools and technology.

3. All sorts of data can now be handled effectively.

4. Enhanced security features reduce data vulnerabilities.

5. TCO is optimized, as cloud infrastructure can be autoscaled as per immediate requirement.

As per Figure 3-3, let's briefly discuss the following Azure components of a modern data warehouse:

1. Azure Data Factory

2. Azure Data Lake Storage

3. Azure Databricks/Spark

4. Azure Synapse Analytics

5. Azure Analysis Services

6. Power BI

**Azure Data Factory** – Azure Data Factory is the orchestration engine available on Azure, which helps to write data pipelines to fetch data from different applications and write to Azure Data Lake Storage or other storage platforms.

**Azure Data Lake Storage** – Azure Data Lake Storage is available storage on Microsoft Azure for big data analytics workloads.

**Azure Databricks or Spark** – Azure Databricks and Spark on HDInsight clusters are managed clusters available on Azure. Having managed services under this category helps to reduce the burden of managing the underlying infrastructure for these complex solutions. Setting up these resources is easy and can be completed within minutes.

**Azure Synapse Analytics** – Azure Synapse Analytics is a new offering available on Microsoft Azure. It's a combination of SQL Data warehouse (MPP offering), Apache Spark, pipelines, and a workspace to manage this entire ecosystem. This is another managed service on Microsoft Azure. As mentioned in Figure 3-3, it can fetch data from Azure Data Lake Storage using PolyBase or data can be directly landed into this platform.

**Azure Analysis Services** – Azure Analysis Services acts as a caching layer for KPIs and all the data needed for dashboards built in Power BI or other tools. It can save lots of money for organizations, as dashboards are the primary mode of data consumption. The rest of the underlying ecosystem can be spun on-demand to refresh this cache.

**Power BI** – Power BI is a managed service to build and publish dashboards. This tool has dozens of connectors to fetch data from various data sources, and the capability to include R/Python to build predictions in the tool itself makes it a great choice for data analytics solutions.

These are the common components that are used for advanced analytics solutions as well. In advanced analytics solutions there are additional components, which are discussed in the next section.

### Advanced analytics on big data

Advanced analytics is the infusion of real-time analytics, modern data warehouse, and machine learning technologies with a focus on predicting future events and forecasting trends. As shown in Figure 3-4, advanced analytics architecture consists of real-time analytics and ML technologies in conjunction with big data technologies.

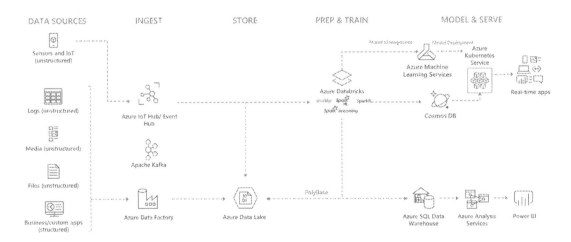

***Figure 3-4.*** *Advanced analytics engine*

Let's briefly discuss the following components in this architecture:

1.  Real-time analytics

2.  Machine learning

**Real-time Analytics**

Real-time analytics is the solution that can consume, analyze, and generate alerts in real time. The data could be captured from any IOT device, clickstream logs, geospatial data, or point of sale (POS) device data, etc.

It includes a real-time analytical and complex event processing engine that is designed to process a high volume of data streamed at a very high speed from a large number of streaming data sources simultaneously. This engine can help in identifying the patterns from these sources, which can further be used to invoke workflows, alerts, or simply archiving the transformed data in storage devices.

The scenarios for real-time analytics include:

- Telemetry data analysis from IOT devices

- Analysis of clickstream data from websites

- Predictive maintenance of hardware infrastructure in plants

- Geospatial analytics for vehicles

- Real-time fraud detection and prevention

- Real-time analytics for inventory management

- Anomaly detection

As shown in Figure 3-4, real-time analytics on Azure uses the following services:

1. Real-time data ingestion services—Event Hubs, Azure IOT Hubs, and Apache Kafka

2. Azure Data Lake Storage

3. Apache Spark Streaming on Azure Databricks or HDInsight

4. Azure Stream Analytics

5. Azure Synapse Analytics

6. Cosmos DB

7. Power BI

1.  **Real-time data ingestion services**: Ingesting the live streaming data from multiple sources using Event Hub or Apache Kafka cluster to Apache Spark streaming or Azure stream analytics. These are all managed services that can be spun within minutes. Security, monitoring, and autoscaling are natively built into these services. These services can ingest data into Azure data Lake Storage using features like event hubs capture or Kafka connect. This data can be used to train ML models or for batch mode data processing.

2.  **Spark Streaming on Azure Databricks**: Azure Databricks is a Spark-based distributed processing engine to process streaming data in real time. Azure Databricks can also be used to combine the real-time streaming data with historic data that is stored in Azure Data Lake Storage or Azure Date warehouse. Azure Databricks also provides a Notebook interface where data scientists can easily write Python, R, or Scala code for creating ML models for deeper analysis and to derive insights of the data.

3.  **Azure Stream Analytics**: Azure Stream Analytics is a serverless real-time stream processing engine. This engine can be used to merge reference data and real-time data to draw analytics in real time.

4.  **Azure Cosmos DB**: For feeding this data to apps in real time, Azure Cosmos DB (which is a multimodel globally distributed NoSQL database), can be utilized. Cosmos DB HTAP capabilities make it a great choice for real-time data to land in Cosmos DB directly.

Synapse Analytics, Azure Data Lake Storage, and Power BI have been discussed under modern data warehouse.

When machine learning is applied on both batch mode and real-time data in a data analytics solution, it becomes advanced analytics. Let's briefly discuss what is machine learning in the architecture mentioned in Figure 3-4.

**Machine Learning**

Machine learning is the process of building algorithms that improve automatically through experience. There are multiple ML methods and algorithms that can be used to address the different types of problems. However, ML in Figure 3-4 refers to various platforms like Azure Machine Learning service, Azure Machine Learning designer service, Spark Mlib, or cognitive services, etc. All these services are available on Azure and can be used as per the complexity of use case.

# Conclusion

In the first three chapters, the major focus was on the key concepts of building data analytics and common design patterns on Microsoft Azure. By now, there should be a clear understanding of data lakes, data warehousing, the advantages of a public cloud, and common design patterns for data analytics solutions. The coming chapters pick up each phase—Data Ingestion, Data Store, Prep and Train, and Model and Serve—from the architectures and discuss them in depth.

# CHAPTER 4

# Data Ingestion

Data ingestion is the first phase of data analytics solutions. In the last three chapters, the major focus was on the basic concepts of data analytics. So far there should be a clear understanding of structured, semistructured, and unstructured data and how the transition of data analytics solutions happened from enterprise data warehouses and big data analytics to modern data warehouses and advanced data analytics. Moreover, overview of all the phases of data analytics solutions has been discussed briefly in the last chapters. In this chapter, the focus will be on the ingest phase and technologies available on Microsoft Azure to build the same.

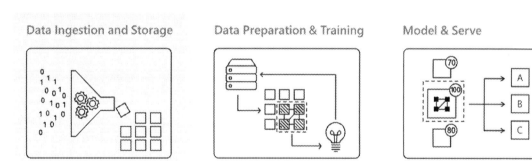

***Figure 4-1.*** *Data ingestion layer*

## Data Ingestion

This is the first phase of data analytics, and it integrates with multiple data sources to bring data into the data analytics engine. As shown in Figure 4-2, the data can come from business applications, customer applications, and streaming applications and IoT sensors.

© Harsh Chawla and Pankaj Khattar 2020
H. Chawla and P. Khattar, *Data Lake Analytics on Microsoft Azure*,
https://doi.org/10.1007/978-1-4842-6252-8_4

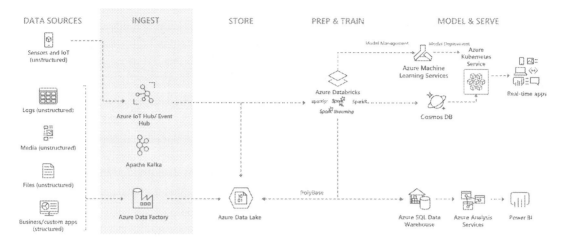

**Figure 4-2.** *Data ingestion phase*

Data ingestion can be categorized in two modes:

1. Real-time mode

2. Batch mode

The journey of data in the data analytics solutions depends on the incoming mode of data. Moreover, the technology stack for both real-time and batch mode is completely different.

# Real-time mode

Real-time data comes from the data source(s) like IoT sensors, streaming applications like clickstream, social media, etc. The typical real-time analytics scenarios built on this data are social media analytics, vehicle maintenance alerts, generating SOS alerts for passengers, or generating alarms depending on the temperature in scenarios like mining and oil exploration, etc. Generally, in these scenarios data is not only processed in real time, but it's also stored in the data storage layer to perform batch processing. This helps to extract information like trends and volumetric data, and generate feedback to improve the system and improve ML models' accuracy, etc.

For real-time data ingestion, since the velocity of incoming data will be high, the solution should have a queuing mechanism to avoid losing any of the event data. Moreover, it should have the capability of connecting to millions of devices at the same time, scale infinitely to handle the load, and process the data in real-time. These factors can enable the solution to have the capabilities to perform real-time operations (e.g, in oil exploration or mining scenarios, if the alert of a specific parameter like rise in temperature or some leakage is reported in real time, it can help to avert accidents). These are scenarios where any delay is not affordable; otherwise the solution will lose its purpose.

Batch mode analysis acts as a feedback system, and it can also help to integrate this data with the data coming from other LOB applications to build 360-degree scenarios. In this section, the discussion is on the technologies shown in Figure 4-3.

***Figure 4-3.*** *Data ingestion technologies*

There are both PaaS and IaaS options available on Azure to process real-time data. However, this section discusses the following managed services:

1. Apache Kafka on HDInsight Cluster

2. Azure Event Hub

3. Azure IoT Hub

There are various advantages to using PaaS services on Azure. Their setup/installation, high availability, monitoring, version upgrades, security patches, and scalability are managed by Microsoft. Every service has multiple tiers and SLAs. An appropriate tier can be chosen, depending on the performance and availability requirement.

Before getting into the details of these services, let's briefly discuss minute yet important concepts of events and messages. These are commonly used terms and their meaning and applicability is summarized in Table 4-1.

**Events** – Events are lightweight notifications of a condition or a state change. The publisher of these events has no expectation about how the event should be handled. However, the consumer of the event decides how this even should be consumed. Events are of two types: discrete and continuous series of values.

**Discrete events** – These events report state change. The consumer of the event only needs to know that something happened. For example, an event notifies consumers that an order was placed. It may have general information about the order, but it may not have details of the order. The consumer may just need to send an email to the customer with confirmation of the order placed. Discrete events are ideal for serverless solutions that need to scale.

**Series events** – These events report conditions that are time-ordered and interrelated. The consumers of these events need the sequenced series of events to analyze what happened. This generally refers to scenarios of data coming from telemetry from IoT devices or distributed streams of data from social media applications.

**Message** - Messages contain the data that triggered the message pipeline. The publisher of the message has an expectation about how the consumer handles the message. A contract exists between the two sides. This option is used in microservices applications for interservice communication.

***Table 4-1.*** *Comparison Table for Events and Messages*

| Service | Purpose | Type | When to use |
|---|---|---|---|
| Event Grid | Reactive programming | Event distribution (discrete) | React to status changes |
| Event Hubs/ Apache Kafka/ IoT Hub, etc. | Big data pipeline | Event streaming (series) | Telemetry and distributed data streaming |
| Service bus | High-value enterprise messaging | Message | Order processing and financial transactions |

In this chapter, the focus is on data coming from telemetry or distributed data streams, and that falls under the series events category. Now, let's discuss real-time data ingestion services in the next section.

# Apache Kafka on HDInsight Cluster

Apache Kafka is one of the most popular solutions for ingesting streaming data. However, not only ingesting streaming data but integrating multiple disparate applications within the organization can also be done through Apache Kafka. In short, anything to do with events and messages can be managed with Apache Kafka—for example, collecting application logs, activity tracking, and gathering data from websites or data from millions of IoT devices.

To cater to the requirement for processing real-time data, this solution is highly efficient, and it can process millions of messages per second. In fact, this platform was built by LinkedIn for purposes like improving user experience or analytics like page views, which content was popular, or recommendation engines. This platform was later donated to the open source community. There are many large companies like Netflix, Twitter, Airbnb, Uber, etc. that use this solution.

Let's discuss the components of the Apache Kafka ecosystem and why it's a scalable and reliable solution for such scenarios. There are five major components in the Apache ecosystem:

1. Producer

2. Consumer

3. Broker

4. Topic

5. ZooKeeper

**Producer** – The producer is the source that produces the message. The producer can be a website generating streams of clickstream data or a fleet truck sharing its location or any IoT sensors producing data streams. The producer just connects to a broker and sends the event without worrying about how it will be processed.

**Broker** – The broker is the server that is part of the Kafka cluster. In other words, a collection of brokers forms a Kafka cluster. Its primary role is to pick up the event from the producer and deliver it to the consumers. Kafka brokers work in leader and worker topology, and there is at least one leader and two in-sync replicas. When an Apache Kafka on HDInsight cluster is built, there are by default two master nodes where the Hadoop ecosystem runs and three mandatory worker nodes (broker nodes) that perform the task.

**Topic** – The topic, like a table in RDBMS, is where the stream of data is written inside apache Kafka brokers. Topics are further divided into partitions, and each message in the partition will be assigned an offset. Offset in a partition becomes the pointer for a consumer to read and process the data.

**ZooKeeper** – ZooKeeper is another component in Apache Kafka, which manages all the brokers starting from replication, availability, and fault tolerance, etc. ZooKeeper has separate servers to manage the Apache Kafka ecosystem; it also needs an odd number of servers, with a minimum of three. Kafka can't run without ZooKeeper. ZooKeeper even works in leader and follower mode, where at least one node will act as a leader and at least two will be followers.

**Consumer** – The consumer is the destination application that processes the event data. In the case of real-time analytics, the same event can be consumed by multiple consumers. In the case of Lambda architecture, when both real-time and batch mode processing is needed, the same event is both passed to stream processing services like Azure Stream analytics, Apache Storm and Spark streaming, etc. and written to Azure Data Lake or Blob storage at the same time for processing later.

Now that we have built some background on Apache Kafka, let's discuss a little bit about how to set up an Apache Kafka on HDInsight cluster. Since detailed documentation is already available on the Microsoft website, we will discuss it on a very basic level.

Steps to set up an Apache Kafka on HDInsight Cluster

1. First, open the Azure portal with the link – `https://portal.azure.com`. Click "create resource" and search for Resource group.

2. Create a resource group and this will hold all the resources for the data analytics solution. All the technology solutions in ingestion, store, prep and train, and model and serve will reside in this resource group.

3. Similarly, create a virtual network, which will host all the resources mentioned in the preceding point.

4. After the resource group and virtual network are created, click "create a resource" to create an HDInsight cluster and enter the details as seen in Figure 4-4.

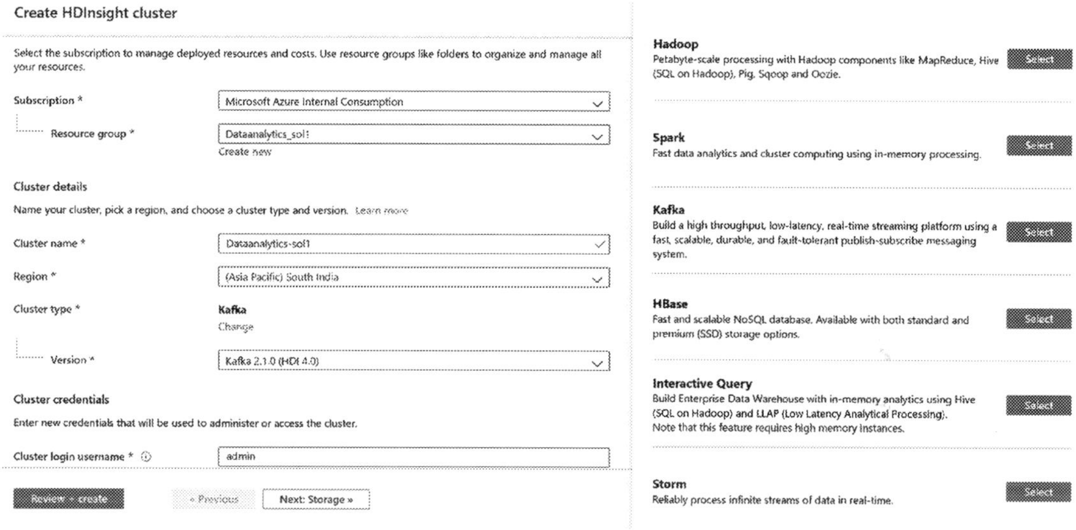

**Create HDInsight cluster**

Select the subscription to manage deployed resources and costs. Use resource groups like folders to organize and manage all your resources.

Subscription *                    Microsoft Azure Internal Consumption

Resource group *                  Dataanalytics_sol1
                                  Create new

**Cluster details**

Name your cluster, pick a region, and choose a cluster type and version. Learn more

Cluster name *                    Dataanalytics-sol1

Region *                          (Asia Pacific) South India

Cluster type *                    Kafka
                                  Change

Version *                         Kafka 2.1.0 (HDI 4.0)

**Cluster credentials**

Enter new credentials that will be used to administer or access the cluster.

Cluster login username * ⓘ        admin

Review + create        « Previous        Next: Storage »

**Hadoop**
Petabyte-scale processing with Hadoop components like MapReduce, Hive (SQL on Hadoop), Pig, Sqoop and Oozie.    Select

**Spark**
Fast data analytics and cluster computing using in-memory processing.    Select

**Kafka**
Build a high throughput, low-latency, real-time streaming platform using a fast, scalable, durable, and fault-tolerant publish-subscribe messaging system.    Select

**HBase**
Fast and scalable NoSQL database. Available with both standard and premium (SSD) storage options.    Select

**Interactive Query**
Build Enterprise Data Warehouse with in-memory analytics using Hive (SQL on Hadoop) and LLAP (Low Latency Analytical Processing). Note that this feature requires high memory instances.    Select

**Storm**
Reliably process infinite streams of data in real-time.    Select

***Figure 4-4.*** *Create HDInsight cluster*

5. Enter the required details; select Kafka from the options on the right side and select the Kafka version (2.1.0 is the latest version available).

6. Enter the password and click "Next Storage," security, and networking details.

7. Following screen (Figure 4-5) is little critical to understand; there is where we need to know the sizing of infrastructure:

49

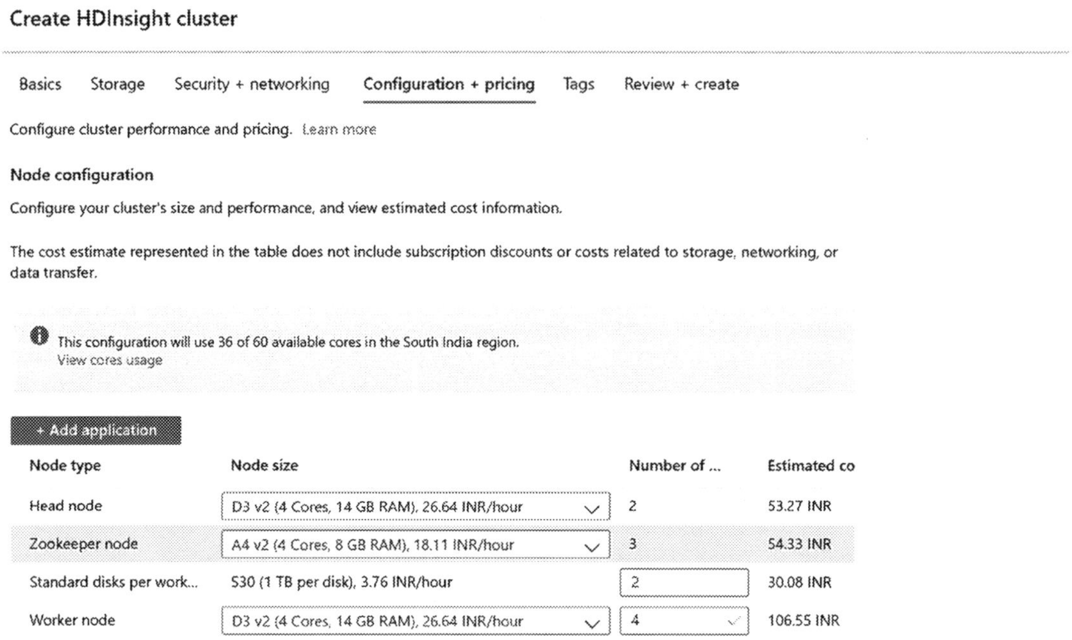

## Create HDInsight cluster

Basics    Storage    Security + networking    **Configuration + pricing**    Tags    Review + create

Configure cluster performance and pricing.  Learn more

### Node configuration

Configure your cluster's size and performance, and view estimated cost information.

The cost estimate represented in the table does not include subscription discounts or costs related to storage, networking, or data transfer.

> ℹ This configuration will use 36 of 60 available cores in the South India region.
> View cores usage

**+ Add application**

| Node type | Node size | Number of ... | Estimated co |
|---|---|---|---|
| Head node | D3 v2 (4 Cores, 14 GB RAM), 26.64 INR/hour | 2 | 53.27 INR |
| Zookeeper node | A4 v2 (4 Cores, 8 GB RAM), 18.11 INR/hour | 3 | 54.33 INR |
| Standard disks per work... | S30 (1 TB per disk), 3.76 INR/hour | 2 | 30.08 INR |
| Worker node | D3 v2 (4 Cores, 14 GB RAM), 26.64 INR/hour | 4 | 106.55 INR |

***Figure 4-5.***  *Enter preferred nodes' size (refer to the sizing section for more details)*

There are four parameters in this screen:

1. *Head node*: There are at minimum two head nodes in an HDInsight cluster; these are the nodes where Hadoop runs along with components like Ambari and Hive Metastore, etc.

2. *ZooKeeper node*: A minimum of three ZooKeeper nodes are required, as discussed earlier; these nodes help to keep the Apache Kafka nodes up and running. Their high availability, replication, and fault tolerance are managed by these nodes.

3. *Standard disk per worker*: A minimum of two disks are required for the setup; however, the number of disks depends on the message size and the retention period.

4. *Worker nodes*: These are the nodes that are called brokers in terms of Apache Kafka. This is where all the topics and partitions are created. A minimum of three worker nodes are needed; however, the number can be increased depending on the expected load.

The cost of the infrastructure is calculated on a per hour basis. With Azure, an Apache Kafka cluster can be up and running within minutes, whereas these steps generally take weeks in an on-premises setup.

**Sizing Apache Kafka on HDInsight clusters** – Now we know about the basic components of Apache Kafka and how it's set up on Microsoft Azure. Let's discuss the critical parameters to consider while sizing the infrastructure for Apache Kafka on HDInsight. There are four typical inputs required to size the cluster:

1. Message rate

2. Message size

3. Replica count

4. Retention policy

Generally, these inputs must be available before the sizing of the cluster. In the case of inputs not being available, Microsoft Azure provides flexibility to scale on demand—it provides flexibility to the architects to start small in the beginning and gradually scale, as needed.

Recommended VM types for Apache Kafka on HDInsight cluster to date are listed in Table 4-2.

**Table 4-2.**  *Recommended VM Types for Apache Kafka*

| Size | vCPU | Memory: GiB | Temp storage (SSD) GiB | Max temp storage throughput: IOPS / Read MBps / Write MBps | Max data disks / throughput: IOPS | Max NICs / Expected network bandwidth (Mbps) |
|---|---|---|---|---|---|---|
| Standard_D3_v2 | 4 | 14 | 200 | 12000 / 187 / 93 | 16 / 16x500 | 4 / 3000 |
| Standard_D4_v2 | 8 | 28 | 400 | 24000 / 375 / 187 | 32 / 32x500 | 8 / 6000 |
| Standard_D5_v2 | 16 | 56 | 800 | 48000 / 750 / 375 | 64 / 64x500 | 8 / 12000 |
| Standard_D12_v2 | 4 | 28 | 200 | 12000 / 187 / 93 | 16 / 16x500 | 4 / 3000 |
| Standard_D13_v2 | 8 | 56 | 400 | 24000 / 375 / 187 | 32 / 32x500 | 8 / 6000 |
| Standard_D14_v2 | 16 | 112 | 800 | 48000 / 750 / 375 | 64 / 64x500 | 8 / 12000 |
| Standard_A1_v2 | 1 | 2 | 10 | 1000 / 20 / 10 | 2 / 2x500 | 2 / 250 |
| Standard_A2_v2 | 2 | 4 | 20 | 2000 / 40 / 20 | 4 / 4x500 | 2 / 500 |
| Standard_A4_v2 | 4 | 8 | 40 | 4000 / 80 / 40 | 8 / 8x500 | 4 / 1000 |

However, the minimum recommended sizes for an Apache Kafka on HDInsight cluster are listed in Table 4-3.

**Table 4-3.**  *Minimum Recommended VM Sizes for Kafka on HDInsight Clusters*

| Cluster type | Kafka |
|---|---|
| Head: default VM size | D3_v2 |
| Head: minimum recommended VM sizes | D3_v2 |
| Worker: default VM size | 4 D12_v2 with 2 S30 disks per broker |
| Worker: minimum recommended VM sizes | D3_v2 |
| ZooKeeper: default VM size | A4_v2 |
| ZooKeeper: minimum recommended VM sizes | A4_v2 |

> **Note**    There is standard Microsoft documentation available on the setup of an Apache Kafka cluster. All the steps can be found on this link: `https://docs. microsoft.com/en-us/azure/hdinsight/kafka/apache-kafka-get- started`

All the preceding information is important to decide which VM will meet the requirement and how many VMs are required to withstand the incoming data streams.

Let's assume the following parameters and arrive at an appropriate sizing:

1. Message rate: 1,000 messages/second

2. Message size: 10 KB

3. Replica count: 3

4. Retention policy: 6 hours

5. Partition: 1

**Step 1** - Total throughput needed would be message rate × message size × replica count.

That is $1,000 \times 10 \times 3 = 30,000 = 30$ MB per second approx.

**Step 2** – Total storage required for this solution is total message size × retention time.

That is 30 MB/s × 6 hours $(6 \times 60 \times 60) = 30 \times 21,600 = 6.4$ TB approx.

Ideal sizing of D3 V2 VM can attach seven 1 TB disks, and a total of 1 VMs will be needed.

After all this exercise and creation of Apache Kafka on HDInsight cluster, it's time for some action. In this chapter, the discussion is only around setting up producers and ensuring the data streams are landing in Apache Kafka. The coming chapters will talk about various options to land this data either in Spark or a storage layer, and how to apply ML for sentiment analytics.

Since an Apache Kafka cluster has been created already, let's configure the producer to pick up live stream data as follows.

# Exercise: Create an Apache Kafka Cluster in Azure HDInsight for Data Ingestion and Analysis

This exercise has the following steps:

1. Create an Apache Kafka cluster on Azure HDInsight

2. Create a topic

3. Ingest data in topic

Since this chapter is about data ingestion, all the exercises are discussed till data ingestion. However, all the remaining phases like event processing and dashboards are discussed in the coming chapters.

Now, let's learn how to create topics and produce events. Since the Apache Kafka cluster has been created in the previous section already, let's log into the cluster to complete the exercise.

Once the cluster is created, we need to connect to the cluster.

1. Run the following ssh command on Windows cmd:

   **ssh** sshuser@CLUSTERNAME-ssh.azurehdinsight.net

   In this command, replace CLUSTERNAME with the name of the cluster.

   When prompted for password, enter the password, and connect to the cluster. After connecting to the Apache Kafka cluster, the first step is to fetch the ZooKeeper and broker information:

   a. Install jq, a command line JSON processor as shown in Figure 4-6.

   ```
   sudo apt -y install jq
   ```

```
sshuser@hn0-kafkac: ~
sshuser@hn0-kafkac:~$ sudo apt -y install jq
Reading package lists... Done
Building dependency tree
Reading state information... Done
The following packages were automatically installed and are no longer required:
  gconf-service gconf-service-backend gconf2 gconf2-common libavahi-glib1 libbonobo2-0 libbonob
  libgnomevfs2-0 libgnomevfs2-common libopts25 liborbit-2-0 libvorbisfile3 linux-azure-cloud-to
  linux-azure-tools-4.15.0-1057 linux-cloud-tools-4.15.0-1057-azure linux-headers-4.15.0-1057-a
  linux-tools-4.15.0-1057-azure sound-theme-freedesktop
Use 'sudo apt autoremove' to remove them.
The following additional packages will be installed:
  libonig2
The following NEW packages will be installed:
  jq libonig2
0 upgraded, 2 newly installed, 0 to remove and 0 not upgraded.
Need to get 231 kB of archives.
After this operation, 797 kB of additional disk space will be used.
Get:1 http://azure.archive.ubuntu.com/ubuntu xenial-updates/universe amd64 libonig2 amd64 5.9.6
Get:2 http://azure.archive.ubuntu.com/ubuntu xenial-updates/universe amd64 jq amd64 1.5+dfsg-1u
Fetched 231 kB in 1s (223 kB/s)
Selecting previously unselected package libonig2:amd64.
(Reading database ... 200582 files and directories currently installed.)
Unpacking libonig2:amd64 (5.9.6-1ubuntu0.1) .................................................
Selecting previously unselected package jq.###############.............................
Preparing to unpack .../jq_1.5+dfsg-1ubuntu0.1_amd64.deb ...
Unpacking jq (1.5+dfsg-1ubuntu0.1) ...###############################....................
Processing triggers for libc-bin (2.23-0ubuntu11) ...####################################
Processing triggers for man-db (2.7.5-1) ...
Setting up libonig2:amd64 (5.9.6-1ubuntu0.1) ...
Setting up jq (1.5+dfsg-1ubuntu0.1) ...############################################
Processing triggers for libc-bin (2.23-0ubuntu11) ...###################################
sshuser@hn0-kafkac:~$
```

***Figure 4-6.*** *Install JQ*

    b. Set up password; replace PASSWORD in the following command:

```
export password='PASSWORD'
```

    c. Export the cluster name:

```
export clusterName='<clustername>'
```

    d. Run the command to extract and set up ZooKeeper information:

```
export KAFKAZKHOSTS=$(curl -sS -u admin:$password
-G https://$clusterName.azurehdinsight.net/api/v1/
clusters/$clusterName/services/ZOOKEEPER/components/
ZOOKEEPER_SERVER | jq -r '["\(.host_components[].
HostRoles.host_name):2181"] | join(",")' | cut -d','
-f1,2);
```

e. Run the command to extract and set up Apache Kafka broker host information.

```
export KAFKABROKERS=$(curl -sS -u admin:$password
-G https://$clusterName.azurehdinsight.net/api/v1/
clusters/$clusterName/services/KAFKA/components/
KAFKA_BROKER | jq -r '["\(.host_components[].
HostRoles.host_name):9092"] | join(",")' | cut -d','
-f1,2);
```

After the preceding configurations are set up, Kafka topics can be created and the events can be produced/consumed.

Kafka comes with an inbuilt utility script named kafka-topics.sh, which can be used to create/delete/list topics. It also provides utility scripts like kafka-console-producer.sh, which can be used to write records to the topics and kafka-console-consumer.sh, which can be used to consume those records from the topic.

Let's run these scripts to manage the Kafka topics:

1. To create a topic test, run the following command:

```
/usr/hdp/current/kafka-broker/bin/kafka-topics.sh
--create --replication-factor 3 --partitions 8 --topic
test --zookeeper $KAFKAZKHOSTS
```

A Kafka topic with replication factor 3 and 8 partitions is created. The worker nodes created are 4; therefore, we can replicate the topics to less than or equal to 4. There are 8 partitions created, which means that the consumer group should have a maximum 8 consumers.

2. To list the topic as shown in Figure 4-7:

```
/usr/hdp/current/kafka-broker/bin/kafka-topics.sh --list
--zookeeper $KAFKAZKHOSTS
```

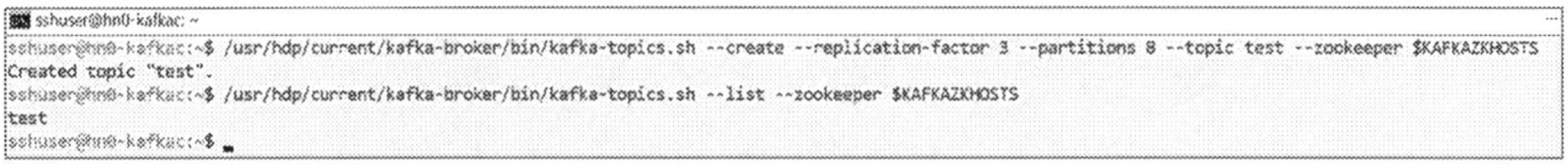

*Figure 4-7.* *Query list of Kafka topics*

3. To write records to topic:

**/usr/hdp/current/kafka-broker/bin/kafka-console-producer.sh --broker-list $KAFKABROKERS --topic test**

4. Type the records as shown in Figure 4-8 and then press Ctl+C to return.

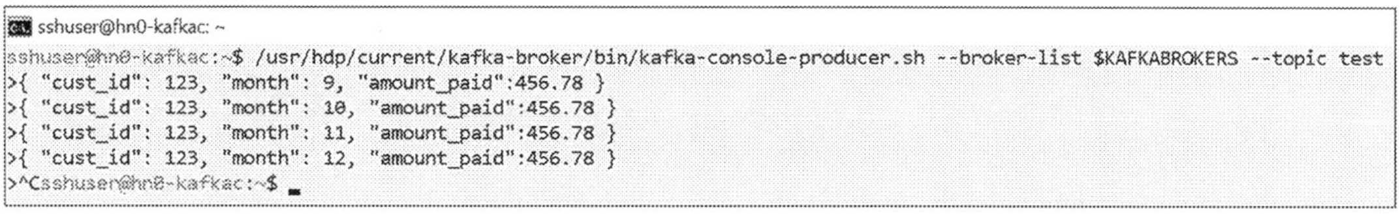

```
sshuser@hn0-kafkac: ~
sshuser@hn0-kafkac:~$ /usr/hdp/current/kafka-broker/bin/kafka-console-producer.sh --broker-list $KAFKABROKERS --topic test
>{ "cust_id": 123, "month": 9, "amount_paid":456.78 }
>{ "cust_id": 123, "month": 10, "amount_paid":456.78 }
>{ "cust_id": 123, "month": 11, "amount_paid":456.78 }
>{ "cust_id": 123, "month": 12, "amount_paid":456.78 }
>^Csshuser@hn0-kafkac:~$
```

***Figure 4-8.***  *Insertion of data streams*

5. To read records from the topic from the beginning:

**/usr/hdp/current/kafka-broker/bin/kafka-console-consumer.sh --bootstrap-server $KAFKABROKERS --topic test --from-beginning**

6. To delete the topic:

**/usr/hdp/current/kafka-broker/bin/kafka-topics.sh --delete --topic test --zookeeper $KAFKAZKHOSTS**

This was a basic exercise to give a hands-on experience on Apache Kafka cluster. Now, let's discuss the next option—Event Hub—as follows.

# Event Hub

Event Hub is a native Azure service for ingesting real-time data streams. This is a turnkey solution based on a serverless offering on Microsoft Azure. There is no need to manage any infrastructure; throughput units can be selected based on the estimated incoming messages. The entire underlying infrastructure is managed by Microsoft. It means there is no need to upgrade or patch, or back up the event service. It reduces maintenance overhead from the infrastructure teams. Refer to Figure 4-9; Event Hub consists of the following components to make it run:

1. Throughput unit

2. Cluster unit

3.  Namespace

4.  Partitions

5.  Event producers

6.  Event receivers

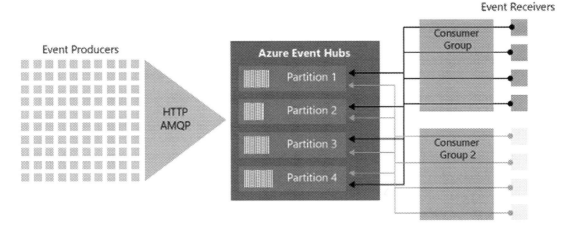

***Figure 4-9.*** *Anatomy of Event Hub ecosystem*

**Throughput unit (TU)** – This is the unit of sizing based on which throughput capacity is defined in Event Hub. Since this is a serverless service, there is no option to select a CPU/memory, etc. Throughput units are translated as follows:

> Ingress: Up to 1 MB per second or 1000 events per second (whichever comes first).
>
> Egress: Up to 2 MB per second or 4096 events per second.
>
> Up to 20 throughput units can be procured in a single namespace. Setup can be started with a minimum of 1 TU, and autoinflate option can be used to increase the TUs up to 20.

**Cluster Unit (CU)** – If the requirement is larger than 20 TU—that is, 20 MBs per second or 20,000 messages per second—the option is to opt for an event hub with dedicated capacity. For dedicated capacity, the measure of throughout is cluster units, and 1 CU = 100 TUs.

**Namespace** – Namespace is a logical collection of event hubs. Throughout units or cluster units are mentioned when the namespace is created. Throughput units/cluster units allocated at the time of namespace creation are shared among all the event hubs.

**Partitions** – Partition is the place where the streaming messages are written by the event producers and are read by the consumers. While creating event hubs, the number of partitions is mentioned. The general recommendation is to create a number of partitions equal to throughput units. The number of partitions should be decided based on the downstream applications.

**Event producers** – Event producers are the entities that produce and send events to event hubs. Some examples of event producers are social media platforms, clickstreams, or application event logs. Event producers write the event to partitions, which are further processed as needed. Event producers can use standard AMQP or HTTP(S) protocols to send data to Event Hub.

**Event receivers** – Event receivers are the entities that consume the events from Event Hub partitions. These are generally applications that process the event data. This can be Apache Spark steaming clusters, Storm, or stream analytics service or even Blob storage. Consumer groups are created for downstream applications. As shown in Figure 4-9, based on the need, consumers can either be part of existing consumer groups or create a new consumer group.

# Why Event Hubs?

After discussing both Event Hubs and Apache Kafka in this chapter, the important question is about making a choice between these two. This is a question that customers very frequently ask. Since Apache Kafka is old in the market, lots of organizations prefer using apache Kafka. However, Azure Event Hubs have evolved a lot due to the following reasons:

1. Microsoft has given lots of flexibility to seamlessly migrate from Apache Kafka to Event Hubs. While creating an Event Hub Namespace, as shown in Figure 4-10, the option to enable Kafka is checked by default.

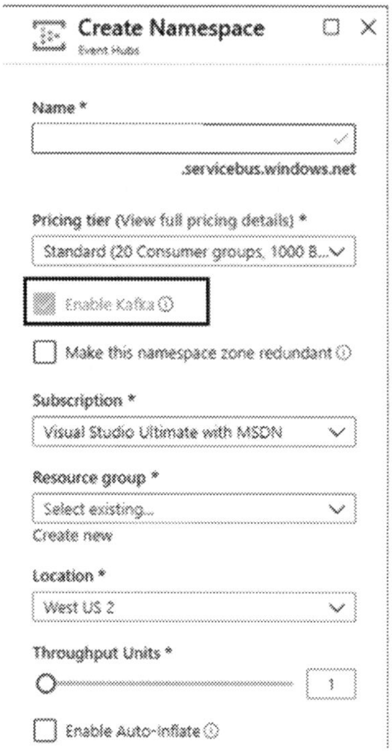

***Figure 4-10.***  *Enable Kafka option checked*

With just a connection string change, event producers can seamlessly integrate with Event Hubs.

2.  *Scalability*: Apache Kafka gives flexibility to pick high configuration VMs, which helps to make the solution highly scalable. Event Hubs even can be deployed on dedicated hosts where high configuration can be picked up, as shown in Table 4-4.

***Table 4-4.*** *Event Hub Service Tiers*

| Feature | Standard | Dedicated |
| --- | --- | --- |
| Bandwidth | 20 TUs (up to 40 TUs) | 20 CUs |
| Namespaces | 1 | 50 per CU |
| Event Hubs | 10 per namespace | 1,000 per namespace |
| Ingress events | Pay per million events | Included |
| Message Size | 1 Million Bytes | 1 Million Bytes |
| Partitions | 32 per Event Hub | 1024 per Event Hub |
| Consumer groups | 20 per Event Hub | No limit per CU, 1,000 per Event Hub |
| Brokered connections | 1,000 included, 5,000 max | 100 K included and max |
| Message Retention | 7 days, 84 GB included per TU | 90 days, 10 TB included per CU |
| Capture | Pay per hour | Included |

Moreover, 1 CU = 100 TU and 1 TU is:

**Ingress**: Up to 1 MB per second or 1,000 events per second (whichever comes first).

**Egress**: Up to 2 MB per second or 4096 events per second.

This means with dedicated hosts, millions of messages per seconds can be easily processed without any maintenance overheads.

Moreover, for parallel real-time, and batch processing, features like event hub capture are helpful. With event hub capture, incoming streaming data can be sent directly to Azure Blob or Azure Data Lake Storage. This feature is like Apache Kafka Connect or Amazon Firehose. For a hands-on experience on Azure Event Hub, let's go through the following exercise.

# Exercise: Ingesting Real-Time Twitter Data Using Event Hub

Today, organizations spend lots of time and money on performing social media analytics to build marketing strategy or improve their service offerings and products. Sentiment analysis is performed on these high-volume Twitter streams of the trending topics. This helps to understand and determine public responses toward the latest products and ideas.

61

In this section, let's discuss how to build a real-time analytics solution that can take continuous streaming data from Twitter into an event hub.

This entire exercise can also be referred to on this link: `https://docs.microsoft.com/en-us/azure/stream-analytics/stream-analytics-twitter-sentiment-analysis-trends`.

This exercise has the following steps:

1. Create Event Hub service

2. A Twitter account

3. Twitter application, which reads the Twitter feeds and pushes them to Event Hub

4. An application that can fetch and send Twitter streams to the Event Hub created in the first step

**Create Event Hub service** - Let's create an event hub namespace and event hub. Event hub namespaces are used to logically group related event bus instances.

1. Create an Event Hub service in the Azure portal.

2. On the Create page, provide the namespace name.

3. Provide details like pricing tier, subscription, resource group, and location for the service.

4. When the namespace is deployed, you can search that in the list of Azure services.

5. From the new namespace, select + Event Hub.

6. Provide the name for the Event Hub.

7. Select Create to create the Event Hub service.

8. Create a shared access signature with manage permission.

After the event hub is created, a Twitter client application needs to be created and configured.

To create a Twitter application, take the following steps:

1. Go to `https://developer.twitter.com/en/apps` and create a developer account. After the developer account is created, there will be an option to create a Twitter app.

2. On the create an application page, provide the required details like purpose to create this application and basic information on the usage of this application.

3. Once the application is created, go to the Keys and Tokens tab. Generate the Access token and Access token secret. Copy the values of Access Token, Access Token Secret, and consumer API keys.

After the event hub and Twitter application are configured, let's make changes in the connection string of the Twitter streams generator application as follows:

1. Download the application code from GitHub: `https://github.com/Azure/azure-stream-analytics/tree/master/DataGenerators/TwitterClientCore`.

2. Use a text editor to open the *App.config* file. Make the following changes to the `<appSettings>` element:

   a. Set `oauth_consumer_key` to the Twitter Consumer Key (API key).

   b. Set `oauth_consumer_secret` to the Twitter Consumer Secret (API secret key).

   c. Set `oauth_token` to the Twitter Access token.

   d. Set `oauth_token_secret` to the Twitter Access token secret.

   e. Set `EventHubNameConnectionString` to the connection string.

   f. Set `EventHubName` to the event hub name (that is the value of the entity path).

3. Open the command line and navigate to the directory where your TwitterClientCore app is located. Use the command `dotnet build` to build the project. Then use the command `dotnet run` to run the app. The app sends Tweets to your event hub.

4. After the Twitter data is ingested, the event hub monitoring dashboard looks as shown in Figure 4-11.

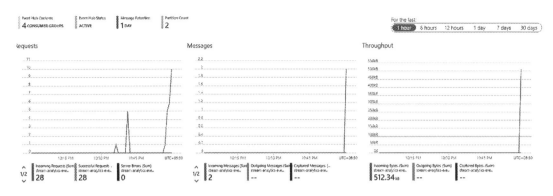

***Figure 4-11.*** *Event hub monitoring dashboard*

It shows the number of incoming requests, messages, and throughput of the event hub. As discussed earlier, event hubs can fetch data from social media, clickstream, or other custom data streaming applications. However, there is another large segment, which generates stream data through IoT devices. Let's discuss IoT Hub, a serverless service available on Azure to connect with these IoT devices and fetch real-time IoT streams.

**IoT Hub –** IoT Hub is a service available on Microsoft Azure for processing data coming particularly from IoT devices. IoT devices are widely used in manufacturing, smart cities, connected cars, or anywhere where real-time data is needed from the hardware devices. The real purpose of IoT is to enhance user experience, detect anomalies, or improve the efficiency of operations without human intervention.

Let's take the example of connected cars: companies use IoT devices to understand the efficiency of the machinery working inside the car and even enhance the user experience. Today, insurance companies use IoT devices to build the driver profile based on their driving behavior and charge them for the insurance accordingly. Even the steering wheel of the car can have sensors like a heartbeat monitor, and this data can be sent to the cloud to offer health alerts, etc.

One caveat with IoT devices is the need to be connected to the Internet to constantly send and receive data from the target cloud server. Consider places like mines, manufacturing units, or where the real-time response is needed and Internet connectivity is low but real-time response is required. Therefore, IoT devices should have the capability to process data on the devices themselves and generate alerts in real-time. Edge devices are IoT devices with a capability to process data locally to a certain extent, generate real-time alerts, and send the output to the cloud whenever there is connectivity.

To connect to IoT devices, Microsoft offers Azure IoT Hub, which are like event hubs but with added features. Both IoT hubs and event hubs are for ingesting real-time data. However, IoT hubs are designed specifically for ingesting data from IoT devices, and event hubs are designed for ingesting telemetry and event data streams. IoT hubs support bidirectional communication between device and the cloud, which helps to do much more with the devices. However, event hubs are unidirectional and just receive the data for further processing. One of the biggest challenges with IoT devices today is the security of data. IoT Hub provides various capabilities like secure communication protocols; integration with security tools on Azure detect and react to the security breaches on IoT devices.

The following chart summarizes the difference between event hubs and IoT hubs tiers.

| IoT Capability | IoT Hub standard tier | Event Hubs |
|---|---|---|
| Device-to-cloud messaging | ✓ | ✓ |
| Protocols: HTTPS, AMQP, AMQP over webSockets | ✓ | ✓ |
| Protocols: MQTT, MQTT over webSockets | ✓ | |
| Per-device identity | ✓ | |
| File upload from devices | ✓ | |
| Device Provisioning Service | ✓ | |
| Cloud-to-device messaging | ✓ | |
| Device twin and device management | ✓ | |
| Device streams (preview) | ✓ | |
| IoT Edge | ✓ | |

To find an updated version of this chart, please check the weblink `https://docs.microsoft.com/en-us/azure/iot-hub/iot-hub-compare-event-hubs`.

In IoT hubs, there are the following important components:

1.  IoT hub units

2.  IoT devices

3.  Partitions

4.  Communication protocols

**IoT hub units** – IoT hub units are the throughput units when creating an IoT hub. There are two tiers, Basic and Standard, available with IoT Hub.

| Tier edition | Sustained throughput | Sustained send rate |
|---|---|---|
| B1, S1 | Up to 1111 KB/minute per unit(1.5 GB/day/unit) | Average of 278 messages/minute per unit(400,000 messages/day per unit) |
| B2, S2 | Up to 16 MB/minute per unit(22.8 GB/day/unit) | Average of 4,167 messages/minute per unit(6 million messages/day per unit) |
| B3, S3 | Up to 814 MB/minute per unit(1144.4 GB/day/unit) | Average of 208,333 messages/minute per unit(300 million messages/day per unit) |

**IoT Devices** – IoT devices must be created to connect to the actual IoT devices. This helps to create two-way communication between IoT Hub and IoT devices.

**Partitions** – The concept of partitions between Event Hub and IoT Hub is the same: stream data is written to partitions. The maximum number of partitions in the IoT hubs can be 32, and the number varies based on the tier. The number of partitions can't be changed after the IoT hub is created.

**Communication protocols** – This is one of the most important concepts for IoT Hub. Security is very important for IoT solutions, and Azure provides a wide range of secure communication protocols. There are three most-used security protocols:

*   *MQTT*: Message Queuing Telemetry Transport

*   *AMQP*: Advanced Message Queuing Protocol

*   *HTTPS*: Hypertext Transfer Protocol Secure

| Protocol | Recommendations |
|---|---|
| MQTTMQTT over WebSocket | Use on all devices that do not require connection to multiple devices (each with its own per-device credentials) over the same TLS connection. |
| AMQPAMQP over WebSocket | Use on field and cloud gateways to take advantage of connection multiplexing across devices. |
| HTTPS | Use for devices that cannot support other protocols. |

Now that the basic concepts of IoT hubs have been discussed, let's get into a basic exercise to learn how to use this service. This exercise has the following steps:

1. Create IoT Hub service

2. Register IoT device

3. Run simulator code to generate input streaming data

4. Query the input data

**Create IoT Hub service**

1. Create an IoT Hub service in the Azure Portal.

2. On the Basic tab, provide details like Subscription, Resource group, region, and IoT Hub name.

3. On the Size and Scale tab, provide details like Pricing and Scale Tier, IoT Hub units, and Azure Security Center.

4. Select Review + create to review and create the instance.

**Register IoT devices**

Once the IoT Hub service is created, the next step is to register a new IoT device in the hub. Let's create a device identity in the identity registry in the IoT hub. A device cannot connect to a hub unless it has an entry in the identity registry.

1. In the IoT Hub service, open IoT Devices, then select New to add a new IoT device.

2. In Create a device, provide a name and choose authentication type and then select Save.

3. Now open the device from the list in the IoT devices pane. Copy the Primary Connection String; this will be used for configuring the IoT device with IoT Hub

The next level is to configure the IoT devices with the IoT Hub.

In order to simulate an IoT device, C#-based simulator code has been created, which is available at:

`https://github.com/Apress/data-lakes-analytics-on-ms-azure/tree/master/`
`AdvanceAnalyticsOnDataLake/Simulator/AdvanceAnalyticsOnDataLake`.

Once downloaded, follow these steps:

1. In the local terminal window, navigate to folder **AdvanceAnalyticsOnDataLake\IOTHub**.

2. Open the file **SensorSimulator.cs** file in the text editor or visual studio, as seen in Figure 4-12.

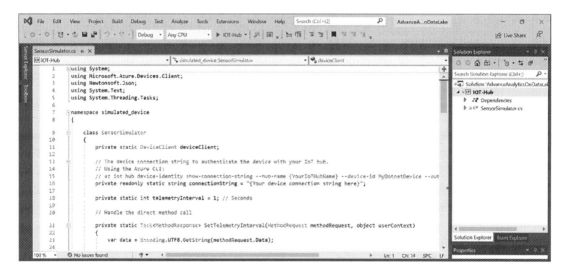

***Figure 4-12.*** *Device connection string*

3. Replace the value of the **connectionString** variable with the device connection string.

4. Run the following command to install the packages in the directory where the code has been downloaded: `dotnet restore`.

5. Run the following command to build the project: dotnet run.

6. After this, the simulator will start sending data to the IoT Hub, as shown in Figure 4-13.

```
Command Prompt
C:\Users\pakhatta\OneDrive - Microsoft\book\project\AdvanceAnalyticsOnDataLake\IOTHub>dotnet run
IoT Hub Quickstarts - Sensor Simulator. Ctrl-C to exit.

4/29/2020 12:43:54 PM > Sending message: {"temperature":25.92357308181169,"humidity":68.23065028909159}
4/29/2020 12:43:55 PM > Sending message: {"temperature":33.35465117513884,"humidity":63.08020052643502}
4/29/2020 12:43:56 PM > Sending message: {"temperature":32.62153322464858,"humidity":79.76162924420164}
4/29/2020 12:43:57 PM > Sending message: {"temperature":21.370967573659016,"humidity":73.31283033514062}
4/29/2020 12:43:58 PM > Sending message: {"temperature":20.924696790019375,"humidity":72.08548414152371}
4/29/2020 12:43:59 PM > Sending message: {"temperature":32.798512996080575,"humidity":71.06894917370236}
4/29/2020 12:44:01 PM > Sending message: {"temperature":22.608426733691445,"humidity":76.50781679735883}
4/29/2020 12:44:02 PM > Sending message: {"temperature":33.370194373824724,"humidity":60.92138170307567}
4/29/2020 12:44:03 PM > Sending message: {"temperature":23.37776152574353,"humidity":79.56789446043219}
4/29/2020 12:44:04 PM > Sending message: {"temperature":23.2354757926592,"humidity":69.91016850336928}
4/29/2020 12:44:05 PM > Sending message: {"temperature":32.882533829138865,"humidity":75.71167184771582}
4/29/2020 12:44:06 PM > Sending message: {"temperature":32.39209075802569,"humidity":79.44645377781544}
4/29/2020 12:44:07 PM > Sending message: {"temperature":34.52844060935473,"humidity":63.58401412311197}
4/29/2020 12:44:08 PM > Sending message: {"temperature":21.463475488807763,"humidity":66.00903494563374}
4/29/2020 12:44:09 PM > Sending message: {"temperature":31.71233422435463,"humidity":66.6269470502748}
4/29/2020 12:44:10 PM > Sending message: {"temperature":23.018577745192953,"humidity":70.14324620838428}
```

***Figure 4-13.*** *IoT simulator run*

Once the devices are configured and the input data ingestion has started, this data can be queried. IoT Hub provides an SQL-like language to fetch information like device twins, message routing, jobs, and module twines.

These queries can be executed in Query explorer, as shown in Figure 4-14.

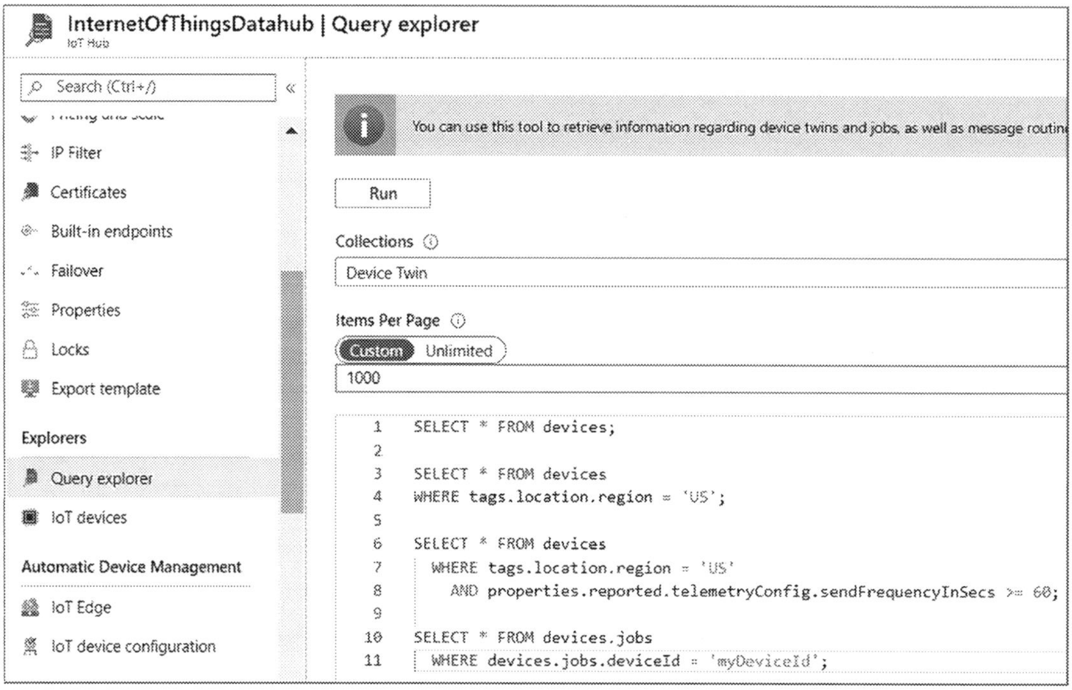

**Figure 4-14.** *IoT Hub query explorer*

For example, the queries could be

```
SELECT * FROM devices;

SELECT * FROM devices
WHERE tags.location.region = 'US';

SELECT * FROM devices
  WHERE tags.location.region = 'US'
    AND properties.reported.telemetryConfig.sendFrequencyInSecs >= 60;

SELECT * FROM devices.jobs
  WHERE devices.jobs.deviceId = 'myDeviceId';
```

With this exercise, real-time data stream ingestion comes to an end. In the next section, the discussion is on batch data ingestion.

# Batch Data Ingestion

IoT hubs and event hubs deal with real-time event streams; however, batch mode data processing is done on periodic intervals. Data coming from various data sources—real-time or batch mode—is stored in a storage solution like Azure Data Lake Storage, Blob storage, or even relational/nonrelations databases. A typical architecture pattern for solutions today is as shown in Figure 4-15 (also refer to Figure 4-1).

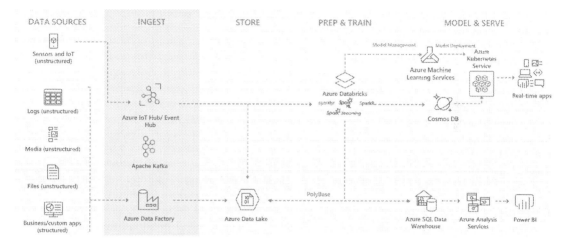

***Figure 4-15.*** *Data ingestion architecture*

To detect patterns or build 360-degree analytics solutions, batch mode processing plays a key role. Real-time processing is required for solutions where mission critical or time-bound response is needed. Examples like fraud detection or predictive maintenance are scenarios where if real-time response is not given, opportunity to avoid an event is lost and can cause financial loss to the business. In fact, batch mode processing acts as a feedback loop that helps to enhance the response for real-time processing; for example, detecting patterns of fraudulent transactions can happen only with batch mode processing, which improves the accuracy of the machine learning model. When the newly trained ML models are used in real time, they can detect frauds/events more effectively.

For batch mode data ingestion, Microsoft Azure offers both PaaS and IaaS solutions. The most widely used solutions on Azure are as follows:

1. Azure Data Factory

2. SQL Server Integration Services on Azure VMs

3. Informatica

4. Azure Marketplace solutions

# Azure Data Factory

Azure Data Factory (ADF) is a Microsoft native cloud solution built for ETL/ELT scenariosc. It's a fully managed PaaS offering from Azure and is a serverless solution. It means data engineers can build data ingestion solution without worrying about the infrastructure requirements. It provides more than 90 data connectors for connecting to multiple on-premises and cloud data sources. Moreover, it provides a code-free user interface where developers can drag and drop activities to create and monitor data pipeline. Data pipeline is another widely used jargon in data analytics solutions.

Data pipeline is the automation of a series of steps that data transverses to yield required output. As we know, the data analytics solution has four major stages: Data Ingestion, Data Storage, Prep and Train, Model and Serve. Solutions like ADF provide the platform to build data pipelines and infuse data into these stages in an orderly fashion.

SQL Server Integration Services and Informatica are the other solutions that have become famous for on-premises environments. These solutions are now cloud ready. Azure Data Factory provides seamless execution of SSIS packages using ADF. Moreover, Informatica is an Azure-certified solution available on Azure Marketplace.

Azure Data Factory provides the following features:

**Connect and collect** – In any data processing tool, the first step is to connect and fetch data from the data sources. These data sources can be databases, file systems, FTP/SFTP web services, or SaaS based systems. It's important that the tool understands the underneath protocol, version, and configurations required to connect to each source. Once the sources are connected, the next step is to fetch data and put that in the centralized location for further processing.

Usually enterprises build or buy data management tools or create custom services that can integrate with these data sources and perform processing. This can be an expensive and time-consuming activity. With the frequent version changes in the source data engines, it's hard to keep pace with regular version upgrades and compatibility checks. Moreover, these custom applications and services often lack features like alerting, monitoring, and other controls that a fully managed PaaS service provides.

All the aforementioned drawbacks are well taken care of in ADF; it provides templates like Copy Activity through which one can easily connect to some 90+ data sources, fetch data, and store in a centralized data store or data lake. For example, data can be collected from an on-premises SQL Server, stored in Azure Data Lake Storage, and further processed in Azure Databricks.

**Transform and enrich** – Azure Data Factory provides features like data flows through which collected data can be easily processed and transformed in the desirable format. Internally, data flows execute on Spark-based distributed in-memory processing, which makes them highly efficient. A developer who is developing data flows need not understand Spark programming for the processing.

Though in case developers prefer to write and execute custom code for transformations, ADF provides external activities that can then execute the custom code on Hadoop, Spark, machine learning, and Data Lake Analytics.

In case there are already predefined SSIS packages available, those can also be executed on the Azure platform with minimal configuration changes.

**CI/CD and publish** – Azure DevOps and GitHub can be easily integrated with Azure Data factory for a full continuous integration and development. It helps in incrementally developing, sharing, reviewing, and testing a process before pushing it to production. By maintaining a code repository, forking branches for multilevel development becomes quick and easy.

**Monitor** – Once the data pipeline building activity is completed, the developer can debug the flow for initial testing and monitoring, once confirmed data flows can be scheduled. For scheduling, ADF provides triggers that can be executed at a specified time or on occurrence of any external event. The external event could be addition or deletion of a file/folder in Azure Data Lake Storage. The developer can also execute pipelines in real time using the trigger row option.

Once executed, the pipelines can be monitored with the status of all the activities included in that pipeline. The performance details are also shared in the monitoring.

It's important to understand the components of the Azure Data Factory toolbox. The following are major components that are used to build pipelines:

1. Datasets

2. Activities

3. Linked service

4. Integration runtime

5.  Pipeline

6.  Data flows

**Datasets** – A dataset is a named reference to the data, which will be used for input or output of an activity. There are more than 90 connectors supported by ADF to create the dataset. The data can be brought from open source DBs, SaaS applications, or any public cloud storage, etc.

**Activities** – Actions to be performed on the data are called activities. Activities take inputs, transform input data, and produce outputs. The following activities are available on ADF:

1.  Move and Transform

2.  Azure Data Explorer

3.  Azure Function

4.  Batch Service

5.  Databricks

6.  Data Lake Analytics

7.  General

8.  HDInsight

9.  Iteration and Conditionals

10. Machine Learning

**Linked service** – Linked service is like a connection string to the data set. There are two types of linked services

1.  *Data Linked Services*: It's the connection to various data sources that ADF can connect with. Currently ADF can connect with more than 90 data sources.

2.  *Compute Linked Services*: It's the connection to various compute resources where the activities mentioned in the above section.

**Integration runtime** – Integration runtime is the infrastructure used by ADF to run the operations like data movement, activity dispatch, data flows, and SSIS package execution. It acts as a bridge between activity and linked service.

**Pipeline** – It's the logical group of activities to be performed on the data. It's a workflow that comes into existence by infusing datasets, linked services, and activities. The relationship between these components can be seen in Figure 4-16.

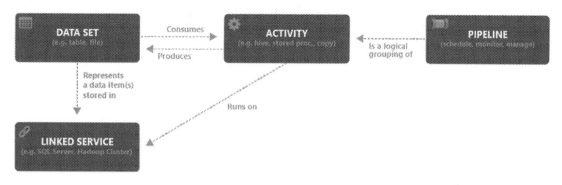

***Figure 4-16.*** *Relationship between ADF components*

**Data Flows** - Data flow is a way to build data pipelines without coding. It's a GUI-based method to create linked services, data sets, and activities and build a complete pipeline. Data flows can be used to create pipelines that are less complex and smaller in size.

# Exercise: Incrementally Load Data from On-premises SQL Server to Azure Blob Storage

In this exercise, we will load the historic, new, and changed data from a table in an SQL Server on-premises instance to Blob storage in an Azure cloud.

A high-level solution diagram looks as shown in Figure 4-17.

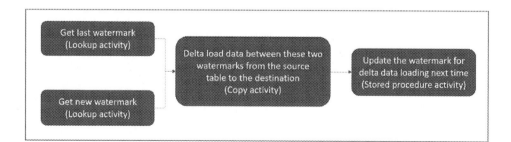

***Figure 4-17.*** *High-level solution diagram*

Important steps include:

1. Select one column in the source table that can be used to identify the new or updated records; normally it is the **last_modify_date_ time** column, which is called the watermark column.

2. Store the latest value of the watermark column in the source column of a separate table, which we will be called the watermark table.

3. Create a pipeline that will have the following activities:

   a. *Lookup activity*: to fetch the last watermark value from the watermark table and fetch the latest watermark value from the source table

   b. *Copy activity*: copies the new and updated data from the source table in SQL Server to Azure Blob storage. The data fetched is greater than the old watermark value and less than the latest watermark value.

   c. *Stored procedure activity*: updates the watermark value in the watermark table with the latest value

Let's start with the exercise:

1. Open the SQL Server Management Studio. In the Object Explore, right-click the database and choose New Query.

2. Create the data source table named **data_source**.

```
create table data_source
(
     PersonID int,
     Name varchar(255),
     LastModifytime datetime
);

INSERT INTO data_source (PersonID, Name, LastModifytime)
VALUES
(1, 'John','5/1/2020 12:56:00 AM'),
(2, 'Bob','5/2/2020 5:23:00 AM'),
```

```
(3, 'Chris','5/3/2020 2:36:00 AM'),
(4, 'David','5/4/2020 3:21:00 AM'),
(5, 'Eger','5/5/2020 8:06:00 AM');
```

3. Create the watermark table to store the watermark value and insert sample rows in it.

```
create table watermarktable
(
TableName varchar(255),
WatermarkValue datetime,
);
INSERT INTO watermarktable
VALUES ('data_source','1/1/2020 12:00:00 AM')
```

4. Create a stored procedure in the SQL database.

```
CREATE PROCEDURE usp_write_watermark @LastModifiedtime datetime, @
TableName varchar(50)
AS
BEGIN
    UPDATE watermarktable
    SET [WatermarkValue] = @LastModifiedtime
    WHERE [TableName] = @TableName
END
```

5. After the SQL setup is done, create Azure Data Factory service on the Azure portal.

6. Enter details: Name, Resource group, and Location and click create.

7. Open ADF and click Author and Monitor.

8. Create an Integration runtime in Azure Data Factory; go to Connections ➤ Integration runtimes as shown in Figure 4-18.

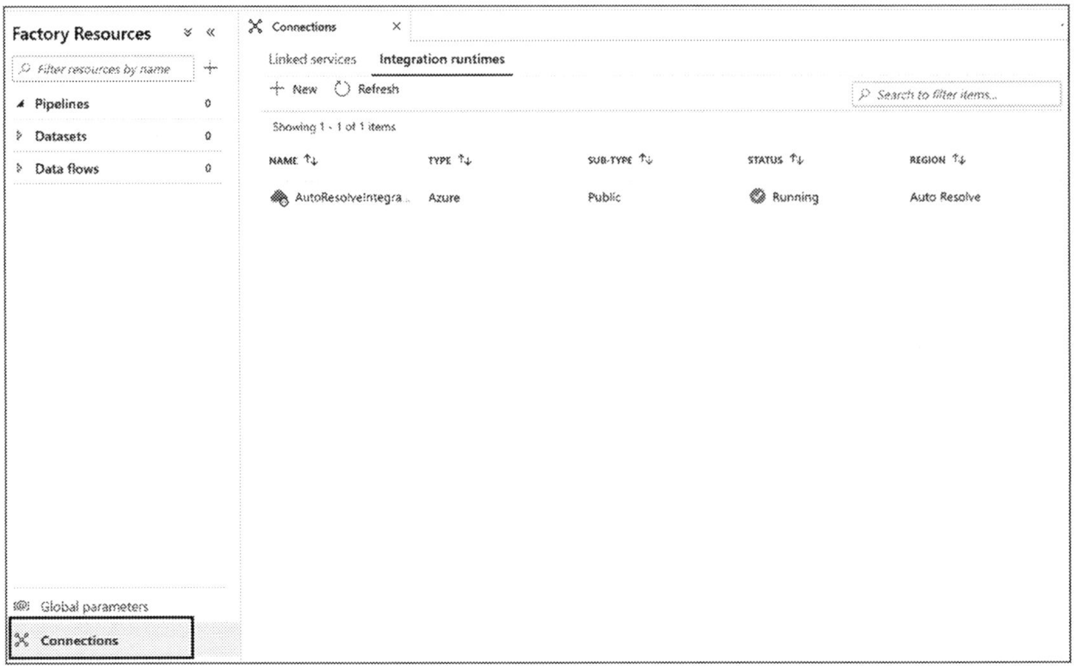

**Figure 4-18.**  *Create integration runtime*

Since the data is being migrated from an on-premises server,
Azure self-Hosted IR is needed. Click New ➤ choose Azure
Self-Hosted ➤ choose Self Hosted ➤ provide name of integration
runtime ➤ click create.

9.  After the Integration runtime is created, download the integration
    runtime and copy the key.

10. Configure the Integration runtime in the on-premises
    environment. Copy the key from step no. 9 and paste in the text
    box and click register.

11. Once setup is complete, it should be connected to the Azure data
    factory and display status as shown in Figure 4-19.

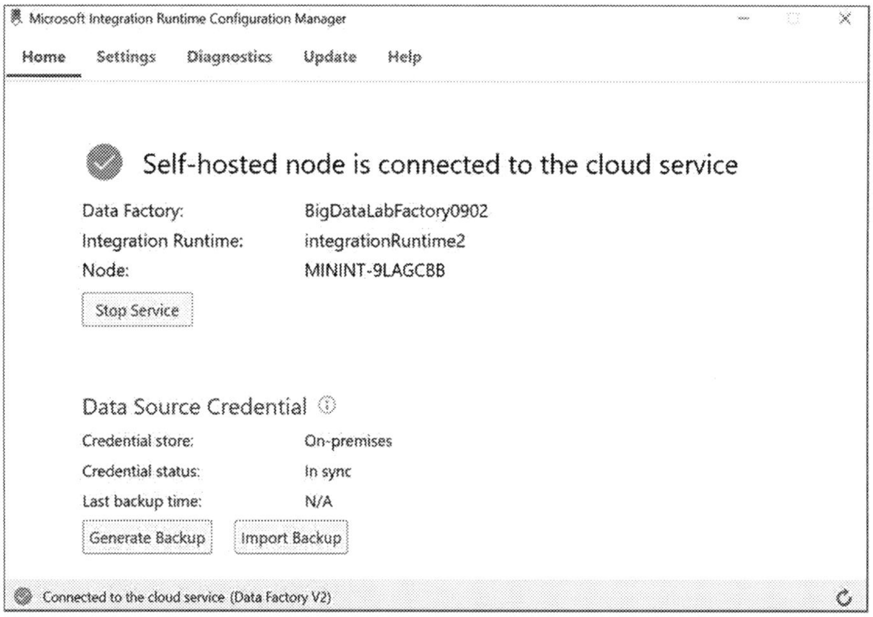

**Figure 4-19.** *Self-hosted runtime connection to ADF*

12. Now that the integration runtime is created, the next step is to create a linked service to the SQL Server instance, as in Figure 4-20.

Go to Connections ➤ Linked Service ➤ New ➤ Select SQL Server ➤ click continue ➤ provide the required details like name, choose integration runtime, provide server name, database name, username, and password ➤ click create.

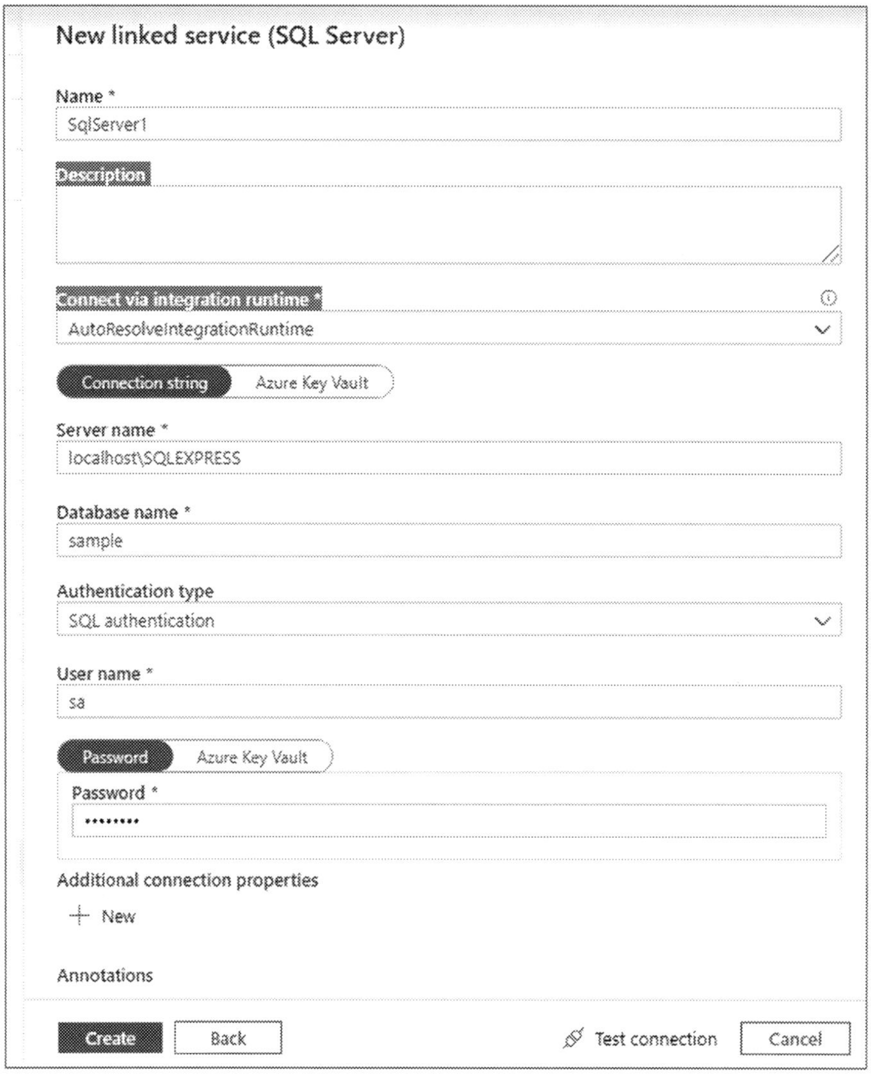

**Name ***

SqlServer1

**Description**

**Connect via integration runtime ***

AutoResolveIntegrationRuntime

**Connection string**    Azure Key Vault

**Server name ***

localhost\SQLEXPRESS

**Database name ***

sample

**Authentication type**

SQL authentication

**User name ***

sa

**Password**    Azure Key Vault

**Password ***

••••••••

**Additional connection properties**

+ New

**Annotations**

**Create**    Back    ⌖ Test connection    Cancel

***Figure 4-20.***  *Local DB connection*

13. Create Azure Blob storage linked service:

    Go to Connections ➤ Linked Service ➤ New ➤ Select Azure
    Blob Storage ➤ click continue ➤ provide the required details like
    name, choose integration runtime, choose azure subscription,
    choose storage account ➤ click create.

14. Now, create a new pipeline in the data factory workspace and
    name it **IncrementalCopyPipeline**.

15. Add the Lookup activity in the pipeline and name it
    **LookupOldWaterMarkActivity.** Switch to the Settings tab and
    click + New for Source Dataset.

    In this step, you create a dataset to represent data in the
    watermark table. This table contains the old watermark that was
    used in the previous copy operation.

16. In the New Dataset window, select Azure SQL Database, and click
    Continue.
    A new window opens for the dataset. In the Set properties window
    for the dataset, enter WatermarkDataset for Name. For Linked
    Service, select the SQL Server Linked service created earlier

17. In the Connection tab, select [dbo].[watermarktable] for Table. If
    you want to preview data in the table, click Preview data, as shown
    in Figure 4-21.

***Figure 4-21.*** *Linked service on ADF for on-prem SQL server*

18. Next, from the Activities toolbox, drag-drop another Lookup
    activity and set the name to **LookupNewWaterMarkActivity** in
    the General tab of the properties window. This Lookup activity
    gets the new latest watermark value from the source table.

19. In the properties window for the second Lookup activity, switch to
    the Settings tab, and click New. You create a dataset to point to the
    source table that contains the new watermark value (maximum
    value of LastModifyTime).

20. In the New Dataset window, select Azure SQL Database, and click Continue. In the Set properties window, enter SourceDataset for Name. Select the SQL Server Linked service. Select [dbo].[data_source] for Table.

21. Now in the activity, choose the settings tab and provide the following query:

```
select MAX(LastModifytime) as NewWatermarkvalue from data_source
```

22. In the Activities toolbox, expand Move and Transform, drag-drop the Copy activity from the Activities toolbox, and set the name to **IncrementalCopyActivity**.

23. Connect both Lookup activities to the Copy activity by dragging the green button attached to the Lookup activities to the Copy activity. Release the mouse button when you see the border color of the Copy activity change to blue, as shown in Figure 4-22.

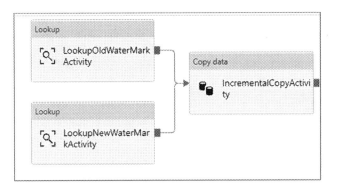

***Figure 4-22.*** *ADF copy activity*

24. Select the Copy activity and confirm that you see the properties for the activity in the Properties window.

25. Switch to the Source tab in the Properties window, and select SourceDataset for the Source Dataset field. Select Query for the Use Query field, and enter the following SQL query for the Query field.

```
select * from data_source_table where LastModifytime > '@{activit
y('LookupOldWaterMarkActivity').output.firstRow.WatermarkValue}'
and LastModifytime <= '@{activity('LookupNewWaterMarkActivity').
output.firstRow.NewWatermarkvalue}'
```

26. Switch to the Sink tab, and click + New for the Sink Dataset field.

27. In this tutorial, sink data store is of type Azure Blob Storage.
    Therefore, select Azure Blob Storage, and click Continue in the
    New Dataset window.

28. In the Select Format window, select the format type of your data,
    and click Continue.

29. In the Set Properties window, enter SinkDataset for Name. For
    Linked Service, select the Linked service created earlier for Azure
    Blob storage, as shown in Figure 4-23

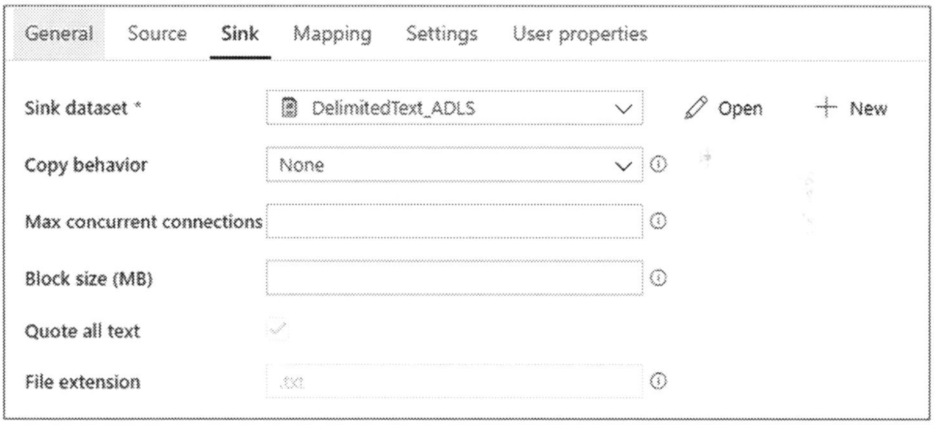

*Figure 4-23.*  *ADLS sink settings*

30. The next step is to drag and drop a stored procedure
    activity, and connect the on-success output of the copy
    activity to stored procedure activity; name the activity
    **StoredProceduretoWriteWatermarkActivity**.

31. In the Settings tab, select the SQL Server linked service; provide
    the SQL server.

When this pipeline is successfully completed, the data from on-premises SQL server will be inserted into Blob storage.

**SQL Server Integration Services –** SQL Server Integration Services is an ETL solution and comes as a part of the SQL Server installation package. It's been famous for building ETL of structured data. There are many organizations that have invested heavily in this technology; when they move to cloud, they can move to Azure Data Factory or may prefer to use SSIS packages only.

Microsoft Azure offers two options to run SSIS packages:

1. Install SSIS on a Windows virtual machine, migrate and run your SSIS packages

2. Run SSIS packages using ADF

The first option is straightforward— it just needs installation of the SSIS component on Azure VM and connection to the SQL database for the SSIS catalog. However, to run SSIS packages using Azure Data Factory, SSIS integration runtime needs to be configured.

Integration runtimes provide the compute infrastructure used by Azure Data Factory for data flow, data movement, activity dispatch, and SSIS package execution.

There are three integration runtimes available on ADF:

1. Azure

2. Self-hosted

3. Azure-SSIS

| IR type | Public network | Private network |
|---------|----------------|-----------------|
| Azure | Data Flow<br>Data movement<br>Activity dispatch | |
| Self-hosted | Data movement<br>Activity dispatch | Data movement<br>Activity dispatch |
| Azure-SSIS | SSIS package execution | SSIS package execution |

**Azure-SSIS IR** – It's the compute instance that can an SSIS package migrated from on-prem or VM installation on a public cloud. It's a fully managed compute that can be scaled up or out depending on the requirement. The SSIS package can be run on this compute with minimal to no change.

# Conclusion

In this chapter, the data ingestion phase for data analytics solutions has been discussed in detail. There are two major modes of data coming into the data analytics pipeline: real-time and batch mode. For both these modes, there are a set of services available on Azure. In this chapter, all these services have been discussed along with exercises for hands on experience. This is a very important chapter for data engineers as they deal with ETL/ELT of incoming data. In the next chapter, the discussion is on how this incoming data can be stored for batch processing.

# CHAPTER 5

# Data Storage

The data store layer plays a vital role in a data analytics solution. This layer stores the data coming from a stream layer or data coming from various applications through an ETL process for further processing. In this chapter, the discussion is around what role the data storage layer in data analytics plays and various storage options available on Microsoft Azure.

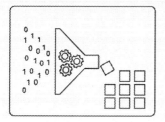
Data Ingestion and Storage

Transfer and store

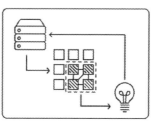
Data Preparation & Training

Process and clean

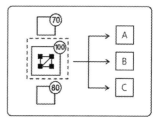
Model & Serve

Serve and analyze

***Figure 5-1.*** *Data storage layer*

## Data Store

This is the transition layer in a data analytics solution. In fact, this is the stage where the transformation journey of data starts. As shown in Figure 5-1 under the Ingest phase, data is coming from event streams for batch mode analysis or from different operational stores through Azure Data Factory (ADF) for further transformation.

© Harsh Chawla and Pankaj Khattar 2020
H. Chawla and P. Khattar, *Data Lake Analytics on Microsoft Azure*,
https://doi.org/10.1007/978-1-4842-6252-8_5

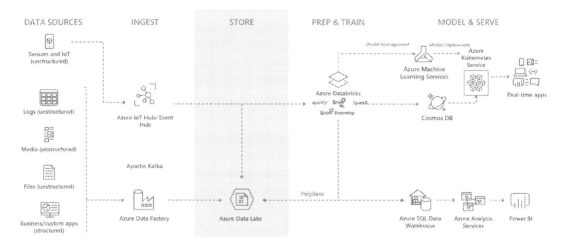

***Figure 5-2.*** *Storage layer in data analytics architecture*

To understand this phase better, let's discuss how having a storage layer saves lots of time and money. Earlier, in typical on-prem enterprise data warehouse scenarios, ETL was done. The following steps were generally taken:

1. Connect to all operational stores through ETL tools.

2. Build centralized operational data stores.

3. Understand KPIs and build data models.

4. Build caching layer using OLAP platforms to precalculate KPIs.

5. Consume these KPIs to build dashboards.

For the structured data stores, the centralized operational data stores were the largest cost. That's because the size of the database instance would be huge, and high-end compute machines were required to process the data. This needed a big upfront investment; that is, both software licenses and hardware had to be bought.

However, in modern data warehouse scenarios, ELT is done instead of ETL. The primary reason for doing that is that public cloud platforms provide a pay as you go buying option. It means the data is brought onto data storage and then this entire data is processed using on-demand compute nodes of preferred technology. In Figure 5-1:

1. Streaming data is landing on Azure Blob Storage for batch processing.

2. Data from structured and unstructured applications is landing on Azure Blob using ADF.

In this case, the data is brought into disks, which is a plain storage method and doesn't need any investment on the software licenses or compute. Under the prep and train phase, data processing services Apache Spark in Azure Synapse analytics can be spun on-demand for a few hours to process this data and move the processed data to the model and serve phase. In this entire process there is no need to make any upfront investment, and only usage-based billing would be paid. Let's discuss the various data store options available on Microsoft Azure in the next section.

# Data Storage Options

For data analytics, there are two major types of storage options preferred on Azure:

1. Blob Storage

2. Azure Data Lake Storage

## Blob Storage

Azure Blob Storage is Microsoft's object storage solution for the cloud. Blob storage is optimized for storing massive amounts of unstructured data. Apart from storing huge amounts of data, it can serve pages to a browser directly, provide distributed access to the files, stream video/audio, and archive data. Blobs/objects in Blob storage can be accessed via Azure Storage REST API, Azure PowerShell, Azure CLI, or an Azure Storage client library/SDKs.

Blob storage is comprised of three types of resources:

- **Storage account** – It provides a unique namespace, and the storage account name must be globally unique. Together with the account name, the container name and object name form the base of addressing every Blob store in Azure Storage uniquely.

- **Container** – Container in the storage account can be considered like a directory to organize your blobs within a storage account.

- **Blob** – Blob in a container can be any unstructured data accessed sequentially. Azure Storage supports three types of blobs:

  - **Block blobs** - Can store text and binary data, up to about 4.7 TB. Block blobs are made up of blocks of data that can be managed individually.

- **Append blobs** - Are made up of blocks like block blobs but are optimized for append operations. Ideal for log files which is optimized for append operations.

- **Page blobs** - Can store files up to 8 TB in size, which can be randomly accessed like a VHD. Page blobs serve as disks for virtual machines.

To select any storage, there are four points that are important to understand:

1. Resilience

2. Data protection

3. Storage lifecycle management

4. Data security

   **Resilience** – To make the storage highly resilient against planned and unplanned events including software maintenance or hardware/network failure, multiple copies of data are stored. In case one copy is not available due to unplanned disruption, the data access requests can be served using other copies. Based on the business need, a customer can segregate data in storage accounts and trade off between cost, high availability, and durability.

   Blob storage supports:

- LRS (locally redundant storage)

- GRS (geo-redundant storage)

- RA-GRS (read-access geo-redundant storage)

   **LRS** - Locally redundant storage copies your data synchronously three times within a single physical location in the primary region. It's the least expensive replication type and offers the least durability when compared with other replications available. LRS provides at least 99.999999999% (11 nines) durability of objects in a year.

**GRS** - Geo-redundant storage copies your data synchronously three times within a single physical location in the primary region using LRS. It then copies data asynchronously to a single physical location in the secondary region. However, data in the secondary location isn't available for read or write access unless there is a failover to the secondary region. You may choose to failover to the secondary region; however, since the nature of replication to secondary region is asynchronous, it may result in data loss. The interval between the most recent writes to the primary region and the last write to the secondary region is known as the recovery point objective (RPO). The RPO indicates the point in time to which data can be recovered. Azure Storage typically has an RPO of less than 15 minutes, although there's currently no SLA on how long it takes to replicate data to the secondary region. You can query the value of the Last Sync Time property using Azure PowerShell, Azure CLI, or one of the Azure Storage client libraries. The Last Sync Time property is a GMT date/time value. GRS offers durability for Azure Storage data objects of at least 99.99999999999999% (16 9s) in a year.

**RA-GRS** – Read-access geo-redundant storage is like GRS and copies your data synchronously three times within a single physical location in the primary region using LRS. It then copies data asynchronously to a single physical location in the secondary region to protect against regional outages. However, the data is available to be read (read only) if the primary region becomes unavailable. RA-GRS offers durability for Azure Storage data objects of at least 99.99999999999999% (16 9s) in a year.

**Data protection** – Data protection deals with methods to save and recover data if it's erroneously modified, corrupted, or deleted.

**Soft delete** - Soft delete protects blob data from being accidentally or erroneously modified or deleted. When soft delete is enabled for a storage account, blobs, blob versions (preview), and snapshots in that storage account may be recovered after they are deleted, within a retention period that you specify. This feature is not yet supported in accounts that have a hierarchical namespace (Azure Data Lake Storage Gen2).

**Blob versioning (preview)** - Previous versions of an object/blob are maintained as specified automatically. You can restore an earlier version of a blob to recover your data if it is erroneously modified or deleted. When blob versioning is enabled, all write or delete operations on block blob automatically trigger a new version except for put block operation. Disabling versioning will not delete the versioned blobs automatically; however, any new versioning of blobs will be disabled.

**Snapshot** - A snapshot is a read-only version of a blob that's taken at a point in time manually when requested by the user.

**Change feed (preview)** - The purpose of the change feed is to provide transaction logs of all the changes that occur to the blobs and the blob metadata in your storage account. The change feed provides an ordered, read-only log of the changes in the blob. This feature is not yet supported in accounts that have a hierarchical namespace (Azure Data Lake Storage Gen2).

**Point-in-time restore for block blobs (preview)** - Point-in-time restore provides protection against accidental deletion or corruption by enabling you to restore block blob data to an earlier state. Point-in-time restore is useful in scenarios where a user or application accidentally deletes data or where an application error corrupts data. Point-in-time restore also enables testing scenarios that require reverting a data set to a known state before running further tests.

**Immutable blobs** – Offers time-based and legal hold policy for storing blobs in WORM (write once, read many) state. Once the policy is set at the container level, and all blobs are updated, users will not be allowed to modify or delete the blobs until the policy expires.

**Storage lifecycle management** – Azure Blob Storage allows you to segregate and store data in the most cost-effective manner. It supports three tiers:

**Hot tier** – For data that needs to be accessed frequently, it offers low access cost but higher storage cost as compared with cool tier. It can be set at storage account as well as blob level.

**Cool tier** – For data that needs to be accessed less frequently as compared with hot tier, hence comparatively high access cost and low storage cost. It can be set at storage account as well as blob level.

**Archive tier** – For data that's infrequently accessed, data will be in offline mode and needs to be rehydrated to hot or cool tier before the user can access them. Comparatively, it offers lowest storage cost but higher access cost including rehydration and access. The archive tier can be set only at a blob level.

**Azure Blob Storage lifecycle management** – Offers a rule-based policy to automatically move data to various tiers and delete the blob at the end of its lifecycle, based on the days specified in the rule. It may be applied to all blobs in a storage account or can be applied to a specific container or type of files.

**Data security** – There are multiple options available to enhance data security, authorization and access control which are mentioned as follows

**Encryption at rest** – Refers to encryption of data in storage accounts using symmetric keys and is by default enabled for all storage accounts. One can chose to use their own keys (customer-managed keys using key vaults) to encrypt data in a storage account.

**Encryption in motion** - Enabling secure transfer for storage account for securing data in transit; makes sure that a call to access data to a storage account is made over HTTPS. It's selected by default for all storage accounts created; the customer can choose to disable it at the time of creation or later.

93

**Access control** – Blobs can be accessed using the Primary/Secondary Access keys, which provide full permission to all the blobs and storage containers in the storage account.

To limit access to the blobs/container/storage account, users can use the following:

**SAS tokens** – Generate SAS (shared access signature) tokens, which can be used to restrict the type of access (read, write, create, delete. etc.) extending to service/container and blobs.

There are three types of SAS tokens:

1.  User delegation SAS - A user delegation SAS token is secured with AD credentials.

2.  Service SAS - Service SAS is generated with access keys and limited to any one of the services like blobs, tables, or files.

3.  Account SAS - An account SAS is secured with the storage account key. An account SAS delegates access to resources in one or more of the storage services.

SAS will have an expiry date, thereby limiting access to the data for a certain duration. SAS tokens can be revoked before their expiry if associated with a stored access policy. The user can also restrict use of tokens to access storage blobs from a specific IP/subnet range.

**Key vault** – Microsoft recommends using Azure AD to authorize requests to Azure Storage. In case shared key authorization is needed, a user may use the key vault. Application can retrieve the keys from the key vault at runtime; hence, sharing the keys with the developers is not needed.

**RBAC** – Role-based access control provides fine-grained access management of Azure resources. It allows applications, users, or groups registered with Azure AD access to resources like storage accounts, virtual machines, virtual networks, etc.

**Enable storage firewall** – Securing storage account access by enabling firewall rules to restrict access from specified IP/subnet ranges or allowing access to trusted Microsoft services only.

**Private endpoint** - A private endpoint assigns a private IP address from a VNet to the storage account. It secures all traffic between VNet and the storage account over a private link.

# Azure Data Lake Storage (ADLS)

In data analytics solutions, data comes from various data sources in disparate formats. The size and scale of data ingestion is large, and there are numerous technologies to process this data (e.g., distributed and MPP systems). There is a need to build a storage platform that can:

1. act as a central repository to break the silos.

2. scale to exabytes and provide GBs/s throughput.

3. seamlessly integrate with a big data analytics ecosystem.

4. provide enterprise level high availability and security.

ADLS is the solution to all the aforementioned needs. It's designed to build a unified solution to store any kind of data. As the name suggests, ADLS can store data coming from various data sources like real-time streaming, structured or unstructured data sources, telemetry, and clickstream logs, etc. ADLS is designed to build Enterprise level big data solutions on Microsoft Azure.

ADLS is built on top of Azure Blob Storage, with capabilities to support big data analytics. ADLS storage has two generations:

1. ADLS Gen1

2. ADLS Gen2

There were a few limitations in ADLS Gen1 that got fixed in Gen2. This chapter specifically discusses ADLS Gen2, as this is the preferred storage option of data analytics workloads.

ADLS supports the following features to make it a great choice for big data workloads:

1. Hierarchical namespace

2. Multiprotocol access

3. Storage lifecycle management

   **Hierarchical namespace**- This has been the most important addition in ADLS gen2. This feature enables ADLS to provide high throughput of object storage with the capability of managing

directories and subdirectories like file storage. This feature
makes ADLS even more suitable for big data analytics workloads.
Without this feature, storage could provide the logical path of the
directories/subdirectories; however, there was no support for
renaming, moving, or deleting the files/directories. The tasks of
renaming, moving, or deleting the directories require movement of
millions of files and that caused lots of contention for big data jobs.

**Multiprotocol access** – Earlier object and analytical storage used
to be managed separately. This further created data silos on Azure
and needed file movement activities at the storage layer to move
the files for analytics operations. With ADLS gen2, this issue has
been resolved. As shown in Figure 5-3, ADLS gen2 supports both
Blob and ADLS APIs.

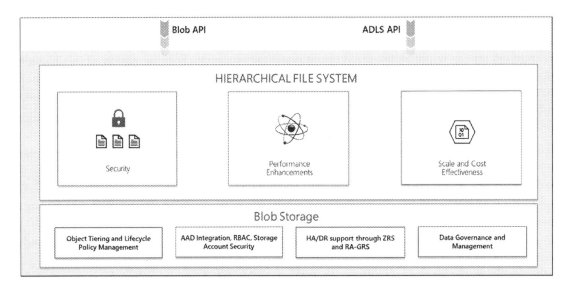

***Figure 5-3.*** *Azure Data Lake Storage architecture*

This means that all the existing big data solutions using Blob API
can seamlessly use ADLS, and newer solutions can use ADLS APIs
for storage operations.

With this multiprotocol access functionality, the following
challenges were resolved:

1. **Azure ecosystem integration** – With the added features of Blob storage and hierarchical namespace, ADLS gen2 seamlessly integrates with the data analytics ecosystem on Azure: solutions like Event Hub, HDInsight clusters, Synapse Analytics, Azure Data Explorer, etc.

2. **Partner ecosystem integration** – Partner solutions like Informatica, Snowflake, and Cloudera, etc. can seamlessly integrate with ADLS Gen2.

This functionality broke silos between Blob and ADLS, and the same application code could function with ADLS seamlessly. ADLS has become the de facto choice for data analytics workloads on Azure.

**Storage lifecycle management** – This is another feature inherited from Blob storage. There are three tiers for the storage lifecycle: hot, cold, and archive. Functionality to automatically switch the tiers of data depending on the duration is now available in ADLS Gen2. Storage lifecycle management is the same as discussed in the preceding section under Blob storage.

Moreover, security, data protection, and resilience features are as mentioned under the Blob section.

# Exercise: Put streaming Data Coming from Event Hubs Directly to Azure Data Lake Storage

To understand ADLS in more detail, let's continue with the Twitter case study mentioned in Chapter 4. As mentioned in Figure 5-2, data from a streaming application can be stored on a storage layer or can be processed with a stream processing solution like Azure Stream Analytics or Spark Streaming.

To do this exercise, follow these steps:

1. Create an Event hub as mentioned in Chapter 4.

2. Run the Twitter stream generator application

3. Enable event capture to send data directly to Azure Data Lake Storage, as shown in the screenshot in Figure 5-4.

97

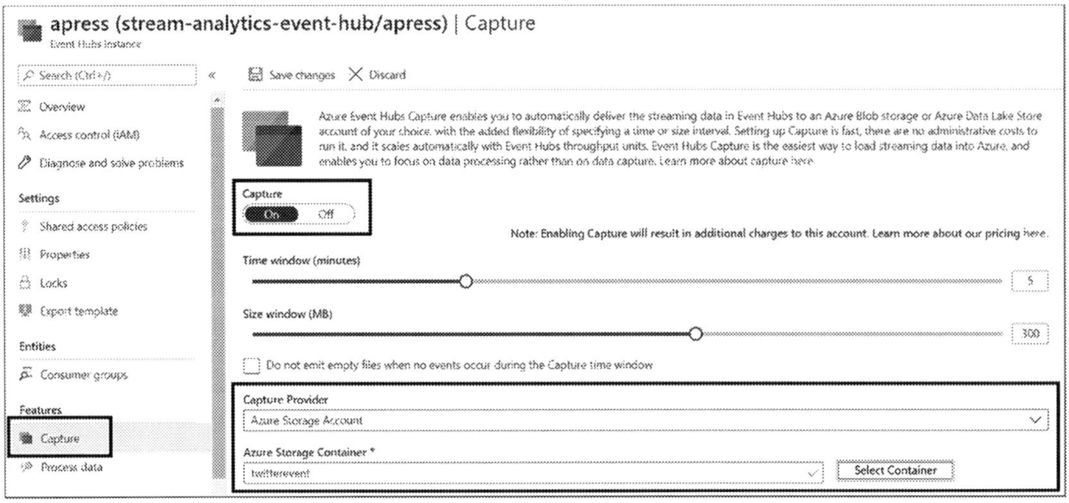

**Figure 5-4.** *Enable event capture in an Event Hub instance*

4. Under Azure Storage, select ADLS Gen2 container

5. To see the data, open storage explorer and connect to ADLS Gen2, as shown in Figure 5-5.

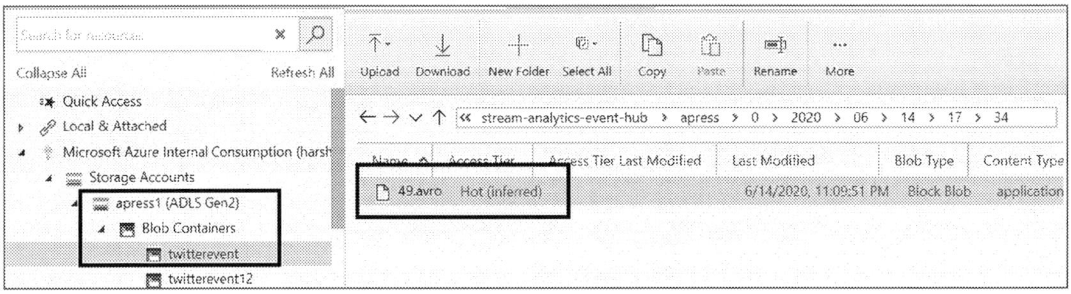

**Figure 5-5.** *Data in Azure Storage*

# Summary

Due to all the new advancements, ADLS gen2 has become a great storage solution for the data analytics ecosystem. As mentioned, data coming from real-time applications in the form of stream or batch mode from disparate data stores can be stored in a single location. From there, it can either be transformed using technologies like Spark or it can move to the model and serve phase, where solutions like Azure Synapse Analytics can access this data for further sharing with downstream applications.

# CHAPTER 6

# Data Preparation and Training Part I

The data preparation and training phase is the most important phase of the data analytics solution. During this phase, data ingested from various sources is merged and crunched together (Figure 6-1). The transformed data further gets infused with machine learning models or is sent to the model and serve phase. The entire data journey is planned, based on the target use case. This phase has been split into two chapters. In this chapter, the discussion is on the various technologies that are applicable in this phase for data analytics. In the next chapter, there are in-depth discussions on advanced data analytics, data science, and various platforms available on Azure to accelerate this journey.

Data Ingestion and Storage          Data Preparation & Training          Model & Serve

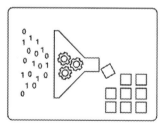
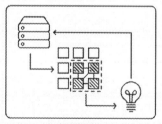
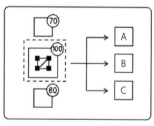

Transfer and store          Process and clean          Serve and analyze

***Figure 6-1.*** *Data processing layer*

H. Chawla and P. Khattar, *Data Lake Analytics on Microsoft Azure*,
https://doi.org/10.1007/978-1-4842-6252-8_6

# Data Preparation and Training

Data preparation and training is the phase where the data (real-time or batch mode) will be processed to extract the desired output. As discussed in the previous chapters, data will come in the form of real-time event streams or it can be picked in the batch mode or it can come from operational databases on an incremental basis. As shown in Figure 6-2, real-time data can either be analyzed in real time and then stored on the cold tier or landed directly on the cold tier, for batch processing. However, batch mode processing can be done independently for data lake and data warehousing scenarios. Till this point, it's been data analytics; when machine learning is introduced to the solution for building predictions and prescriptions, it becomes advanced data analytics.

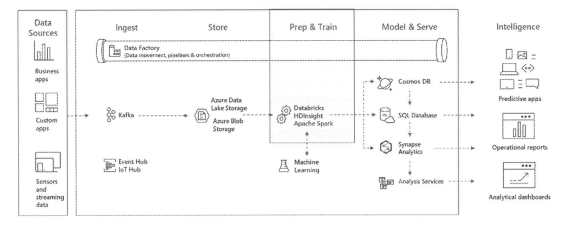

***Figure 6-2.***  *Real-time and batch mode data processing*

The terms ETL and ELT have been popular for decades; however, data preparation/ wrangling are new terms in data analytics concepts. ETL/ELT is generally built by data engineers, based on the requirement from business users. However, data wrangling or preparation is the process to bring data closer to business users. It means that people who want to consume data should be able to prepare and explore the data. Today, there are multiple data sources (e.g., data coming from clickstreams, social media, telemetry, IoT, etc.) of various types' however, ETL is primarily designed for structured data or data coming from operational data stores. Therefore, there are additional steps required to prepare the data for consumption. Data preparation is the process of improving data quality, to make the raw data ready to be analyzed (i.e., use it to train ML models or send it directly to the model and serve phase):

1. *Data cleaning:* Process to detect and remove noisy, inaccurate, and unwanted data

2. *Data transformation:* Normalize data to avoid too many joins and improve consistency and completeness of data

3. *Data reduction:* Data reduction is aggregation of data and identifying data samples needed to train the ML models.

4. *Data discretization:* Process to convert data into right-sized partitions or internals to bring uniformity

5. *Text cleaning:* Process to identify the data based on the target use case, and further cleaning the data that is not needed

Before proceeding, it's worth mentioning here the two personas involved in this phase (Figure 6-3):

1. *Data engineers:* Data engineers manage the entire pipeline of ingestion and storage of data. The focus area for data engineers is to design, build, and arrange data pipelines. They are equipped with skills like advanced programming and analytics, distributed systems, and data pipelines.

2. *Data scientists:* ML developers and data scientists focus on exploratory data analysis and building ML models. They are responsible for creating hypothesis testing, analyzing and building ML models using clean data. They are equipped with skills like ML/AI knowledge, advanced statistics, and advanced analytics.

There are a few sets of skills that are common for both data engineers and data scientists: big data programming and analytics.

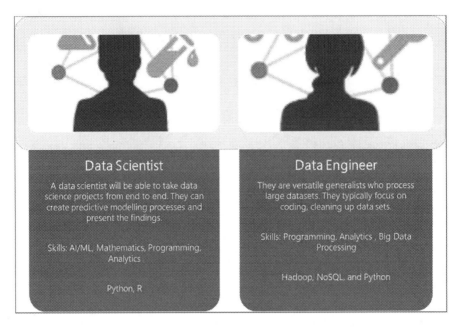

*Figure 6-3.* *Roles of data scientist and data engineer*

Let's further discuss the data preparation and training phase.

# Data Preparation

By now there is a brief understanding of what happens under the data preparation phase in data analytics solutions. Data preparation is part of larger term called data processing. There are two major scenarios to process the data:

1. Process real-time data streams

2. Process batch mode data

## Process Real-Time Data Streams

Real-time data streams can come from platforms like social media, websites, clickstreams, or can be telemetry data as well. To ingest this data on Azure, the following technologies are available:

1. Managed service - PaaS (platform as a Service)

   a. Apache Kafka on HDInsight Cluster

   b. Event Hub

   c. IoT Hub

2. Infrastructure as a service

   d. Apache Kafka

   e. Rabbit MQ

   f. Zero MQ

   g. Marketplace solutions

In Chapter 4, real-time streams and batch data ingestion were discussed in detail. Moreover, there were exercises based on Apache Kafka, Event hubs, and IoT hubs. In the coming sections, let's delve deeper into the prep and train phase and discuss how the data coming from the aforementioned data sources can be processed.

Based on our customer interactions and learning from the field, the following technologies are the most recommended:

1. Apache Spark on HDInsight clusters or Databricks

2. Stream analytics

## Apache Spark

Microsoft Azure has managed Apache Spark on HDInsight cluster and Azure Databricks cluster, which can be spun with a few clicks. Apache Spark can be used for both streaming and batch mode data processing.

Let's delve deeper into the Apache Spark ecosystem and understand the concepts in detail.

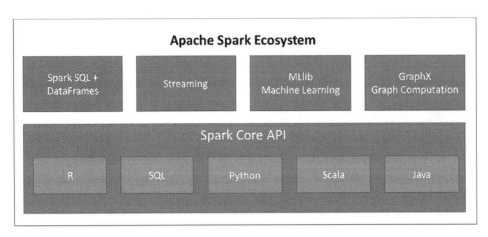

***Figure 6-4.*** *Apache Spark ecosystem*

As shown in Figure 6-4, the Apache Spark ecosystem consists of the following components:

- *Spark DataFrames*: A DataFrame is a distributed collection of data organized into named columns. It is conceptually equivalent to a table in a relational database or a data frame in R/Python. DataFrames can be built from structured data sources, tables in Hive, RDDs, or databases. DataFrames can be constructed in R, Python, Scala, and Java languages.

- *Spark SQL*: For structured data processing, Spark provides Spark SQL, which can be used to execute SQL queries on structured data. Spark SQL provides Spark with more information about the type and structure of the data, which helps Spark to provide query processing optimization as compared with DataFrames.

- *Spark Streaming*: For streaming data processing, Spark provides Spark Streaming. It enables distributed, scalable, fault-tolerant, high-throughput processing of real-time streaming data. Data in this case can be ingested through sources like Kafka, Kinesis, Flume, or TCP sockets. It supports functions like map, reduce, and join window, which can be used for high level processing. Further, processed data can be pushed to databases, filesystems, or BI tools.

- *Spark MLlib libraries*: These are Spark's machine learning libraries, which make machine learning distributed, scalable, and efficient. They provide common algorithms like classification, regression, clustering, and collaborative filtering. Moreover, Spark MLlib has the capability to build complete data science models. Capabilities like features extraction, transformation, and selection of data or even tools for constructing, evaluating, and tuning ML based pipelines are available with Spark MLlib.

- *GraphX*: This component in Spark helps in processing graphs through parallel computation; its use cases include data exploration and cognitive analytics.

- *Spark Core API*: It provides support for languages like Scala, Java, SQL, R, and Python.

Apart from this, there are concepts of Apache Spark APIs that are important to understand:

1. Programming APIs

2. Catalyst optimizer

3. Structured streaming

**Programming APIs**

There are three programing APIs available with Apache Spark:

- Resilient Distributed Dataset

- DataFrame

- Dataset

Let's discuss these APIs, outline their performance benefits, and target use cases.

- *Resilient Distributed Dataset (RDD)*: In the earlier version of Spark, RDD was the primary API. RDD is an immutable distributed data collection, which can be partitioned over the Spark cluster to make data processing run in parallel. Activities like transformation and actions can be performed over RDDs. RDDs are preferably used when low-level access or control is required over the datasets or

when the data is unstructured, like media files or free text data, and
there is no need to impose any schema. Moreover, data manipulation
is to be achieved through pure functional programming.

- *DataFrame*: Just like RDDs, they are also immutable distributed data
collections, but here data is organized into columns just like tables
in relational databases. It gives options to the developers to provide a
structure to the data, thus it has a high-level abstraction. DataFrames
are much simpler to understand and work with, as this API provides
many functions to manipulate the data as well.

- *Dataset*: Dataset is an extension to DataFrame. It has two API
characteristics, which are strongly typed and untyped. By default,
they are strongly typed JVM objects. Datasets provide better
performance and optimization over DataFrames, as they expose
expressions and data fields to a query planner through a catalyst
query optimizer and Tungsten execution engine. Since Spark 2.0,
DataFrame APIs have been merged with Dataset APIs to provide a
unified data processing capability across the multiple Spark libraries.

DataFrame or Dataset should be used when:

1. high-level abstraction and simplified APIs are needed.

2. filters, map, aggregation, sum, or lambda functions on structured
   or semistructured data is needed.

3. type safety at compile time and high level of optimization is
   needed.

**Catalyst optimizer**

DataFrames and Datasets are based on the Spark SQL engine. Spark SQL consists
of a catalyst optimizer, which can be used by both DataFrame and Dataset APIs. The
Catalyst optimizer (Figure 6-5) is an extensible query optimizer; it leverages advanced
programming languages features like Scala's pattern matching and quasiquotes.

Pattern matching is a mechanism to check a value against a pattern; a pattern match
includes a sequence of alternatives, each starting with the keyword case. Each alternative
includes a pattern and one or more expressions, which will be evaluated if the pattern
matches. An arrow symbol => separates the pattern from the expressions.

Quasiquotes are a notation that manipulates syntax trees with ease. In simple terms, quasiquotes are a way to transform a text into executable code. There are multiple advantages of using quasiquotes, as they are type checked at compile time, thus ensuring appropriate ASTs or literals substitution. They return AST and provide compiler optimizations. The optimizations are both rule based and cost based.

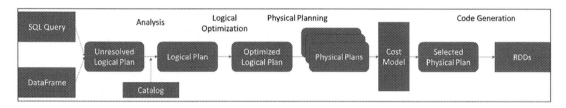

***Figure 6-5.***  *Catalyst optimizer*

### Structured streaming

Apache 1.X version had a concept called micro batching: based on a specific interval, the streaming data will be gathered and processed. It used to create an RDD based on the intervals mentioned for micro batch. Moreover, APIs for streaming data and batch data were not fully compatible.

Spark 2.0 solved the programming challenges and removed the overhead from developers. Spark 2.0 has a concept called DataFrames, which is another layer built on top of RDDs. This version of Spark provides unified access to both streaming and batch data: APIs used for both stream and batch data are the same. Moreover, the data is stored in the form of unbounded data sets. It's called structured streaming (Figure 6-6). Structured streaming is scalable, fault-tolerant, high-level streaming built on Spark SQL engine. It supports batch, interactive, and streaming queries. Spark SQL ensures it runs incrementally and continuously, and it updates the result as streaming data continues to arrive. Internally, by default, structured streaming queries are processed using a micro-batch processing engine, which processes data streams as a series of small batch jobs. It helps to achieve end-to-end latencies as low as 100 milliseconds and guarantees exactly once fault-tolerance.

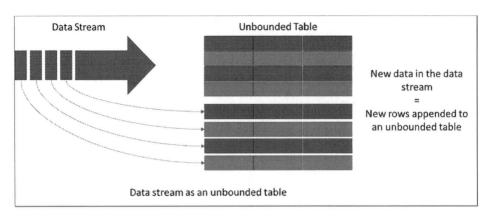

***Figure 6-6.*** *Unbounded table to support structured streaming*

## Continuous Application

Most streaming engines only focus on performing computations on streaming data. There are a few use cases that only involve streaming data, while most include batch data as well. Continuous application (Figure 6-7) is such an application that involves both streaming and batch workloads, including tasks such as:

- Updating backend data in real time

- Extract, transform, load (ETL)

- Machine learning on continuous real-time data

- Building a real-time version of a batch job

Spark supports building continuous applications where the same query can be executed on batch and real-time data, perform ETL operations, and generate reports.

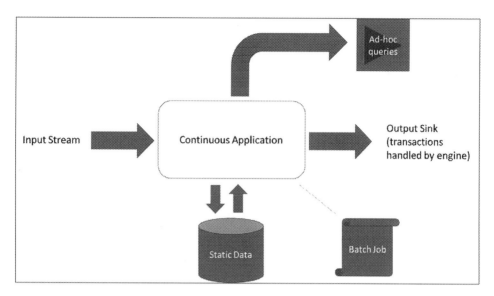

***Figure 6-7.*** *Continuous application reference architecture*

## Apache Spark in Azure Databricks

Azure Databricks (Figure 6-8) is an Apache Spark-based analytics platform that is optimized for Azure Cloud. Databricks is designed by the founders of Apache Spark, and it's integrated into Azure as a first-party service for one-click setup, which streamlines the workflows and provides interactive workspace. This enables collaboration between data engineers, data scientists, and business analysts.

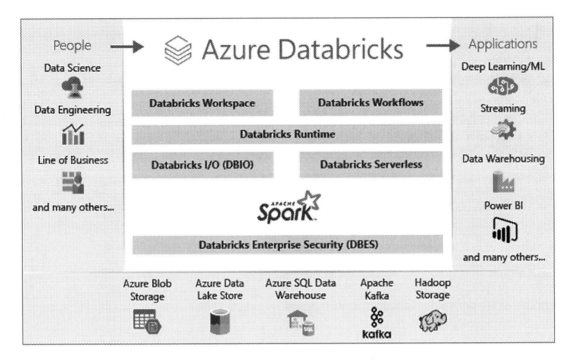

***Figure 6-8.*** *Azure Databricks ecosystem*

With Apache Spark in Azure Databricks, apart from the benefits mentioned, the additional benefits are as follows:

- Fully managed Spark clusters, which are created in seconds

- Clusters can autoscale, depending on the load

- Interactive workspace for exploration and visualization

- A unified platform to host all the Spark-based applications

- Instant access to the latest Spark features released by Databricks

- Secured data integration capabilities built to integrate with major data sources in Azure and external sources

- Integration with BI tools like PowerBI and Azure data storages like SQL Data Warehouse, Cosmos DB, Data Lake Storage, and Blob storage

- Provides enterprise-level security by integrating with Azure Active Directory, role-based access controls (RBACs) for fine-grained user permission, and enterprise-grade SLAs

## Exercise: Sentiment Analysis of Streaming Data Using Azure Databricks

By now, the basic concepts of Apache Spark and Databricks have been discussed. Let's understand how to process streaming data with an Apache Spark Databricks cluster, with the help of an exercise. This exercise is about performing sentiment analysis on the Twitter data using Azure Databricks in real time.

As per the solution architecture in Figure 6-9, the following services will be used:

1. *Event Hub*: for data ingestion

2. *Azure Databricks*: for real-time stream processing

3. *Cognitive services*: for text analytics

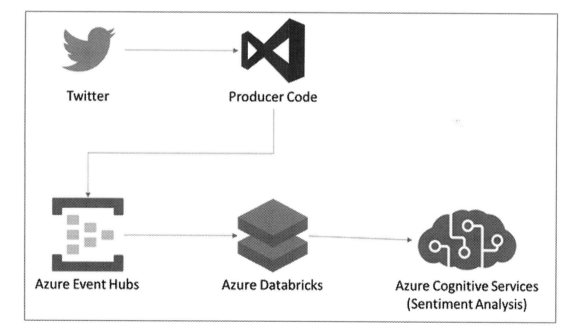

***Figure 6-9.*** *Solution architecture for Twitter sentiment analytics*

The exercise includes the following high-level steps:

1. Create an Azure Databricks workspace.

2. Create a Spark cluster in Azure Databricks.

3. Create a Twitter app to access streaming data.

4. Create an Event Hub instance.

5. Create a client application to fetch Twitter data to push to Event Hub.

6. Read tweets from Event Hubs in Azure Databricks.

7. Perform sentiment analytics using Azure cognitive services API.

**Step 1** - Create an Azure Databricks workspace as mentioned in this article:
`https://docs.microsoft.com/en-us/azure/azure-databricks/quickstart-create-databricks-workspace-portal`.

1. In the Azure portal, create an Azure Databricks workspace by selecting Create a resource ➤ Data + Analytics ➤ Azure Databricks.

2. To create the Azure Databricks service, provide the required details, which are:

   - Workspace name

   - Subscription

   - Resource group

   - Location

   - Pricing tier

3. Select create, and create the Azure Databricks service

**Step 2** - Create a Spark cluster in Databricks as per the steps mentioned in this article: `https://docs.microsoft.com/en-us/azure/databricks/clusters/create`.

1. In the Azure portal, go to the Databricks workspace and select launch workspace.

2. Webpage will be redirected to a new portal; select New Cluster.

3. In the New Cluster page, provide the following details:

    a.  Name of the cluster

    b.  Choose Databricks runtime as 6.0 or higher.

    c.  Select the cluster worker & driver node size from the given list.

    d.  Create the cluster.

**Step 3** - Let's create a Twitter application: This step is already discussed in Chapter 4 in the Event Hub exercise.

1. Go to `https://developer.twitter.com/en/apps` and create a developer account. After the developer account is created, there will be an option to create a Twitter app.

2. On the create an application page, provide the required details like the purpose for creating this application and basic information on the usage of the application (Figure 6-10).

***Figure 6-10.*** *Twitter app details*

3.  Once the application is created, go to the Keys and tokens tab. Generate the Access token and access token secret. Copy the values of customer API keys, Access token, and Access token secret (Figure 6-11).

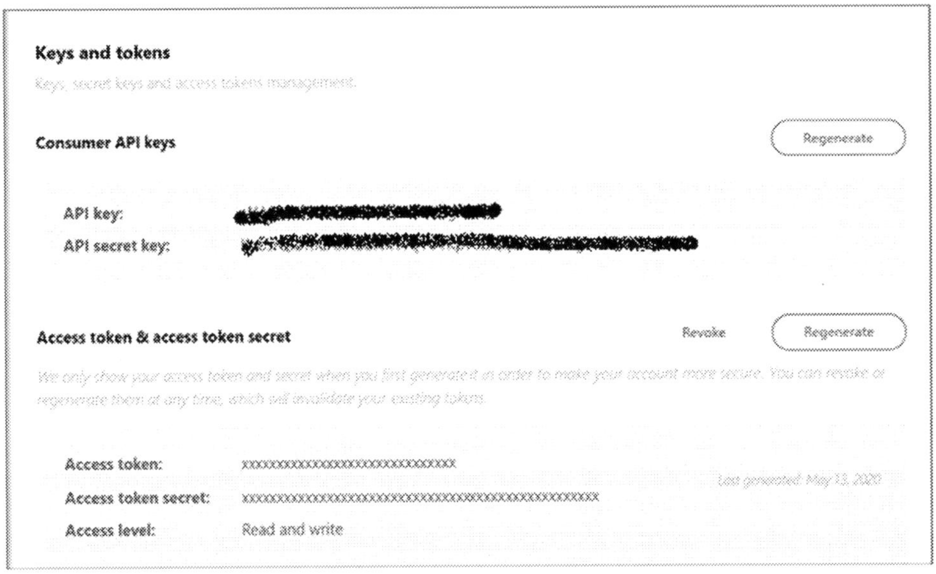

***Figure 6-11.*** *Twitter app keys and tokens*

**Step 4** – Create an Event Hub instance in the Azure portal. This step is already shared in Chapter 4 in the Event Hub exercise. For reference, the following steps need to be followed:

1.  In the Azure portal, select Event Hub and create a Namespace.

2.  Provide the following details (Figure 6-12):

    a.  Select the subscription.

    b.  Select the resource group or create a new resource group.

    c.  Provide the name of the Namespace.

    d.  Select the location of the Namespace.

    e.  Choose the Pricing tier, which could either Basic or Standard.

    f.  Provide the Throughput units settings: default is 1.

    g.  Click the Review + create button.

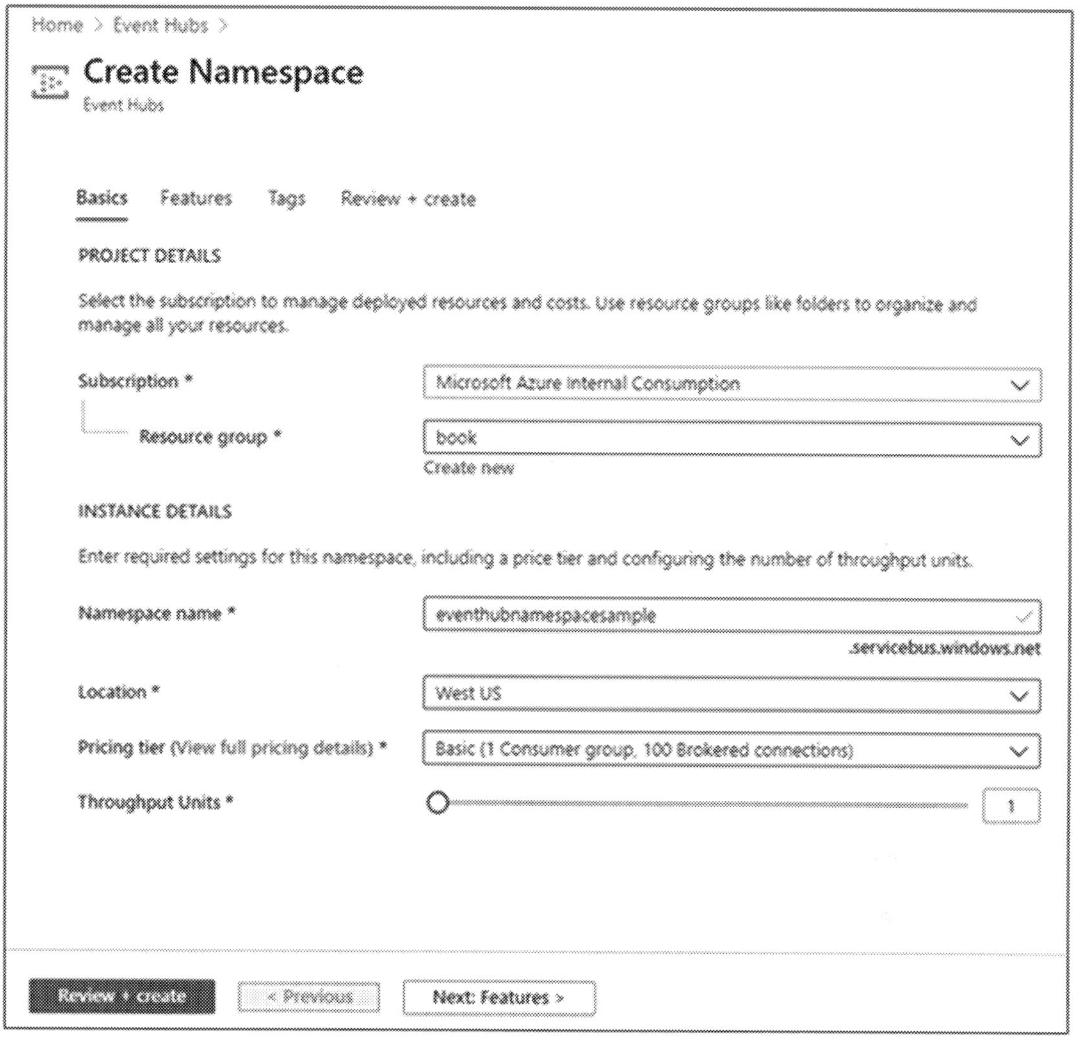

**Figure 6-12.** *Create Event Hub instance*

3. Once the Namespace is created, create an Event Hub:

   a. On the Event Hub namespace, select the Event Hub under Entities.

   b. Click the + button to add new Event Hub

   c. Type a name for the Event Hub and click create.

**Step 5** - Create a client application to fetch data from Twitter and send it to Event Hub:

1.  Create a Java application that will fetch the tweets from Twitter using the Twitter APIs, and then send the tweets to Event Hub.

2.  The sample project is available at GitHub at `https://github.com/Apress/data-lakes-analytics-on-ms-azure/tree/master/Twitter-EventHubApp`.

3.  In the project, Maven-based libraries have been used to connect to Twitter and Event Hub.

4.  A Java class named SendTweetsToEventHub.java has been created; it reads the tweets from Twitter and sends them to Event Hub on a continuous real-time basis.

5.  Refer to the project on GitHub and execute the project after passing the relevant configurations of Twitter and Event Hub (refer to the project file).

6.  Once you run the code, the output displayed will be something like Figure 6-13.

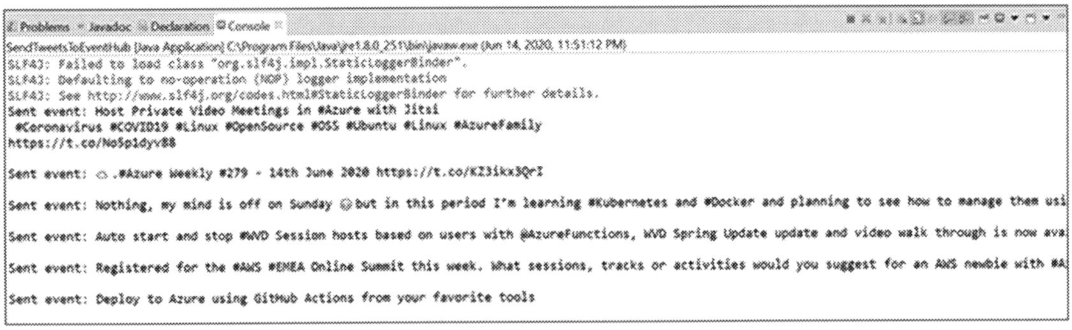

***Figure 6-13.*** *Client application output console*

**Step 6** – Now, let's read data ingested from Event Hub in the Databricks cluster:

1. Install the Event Hub library in the Spark cluster.

2. In the Azure Databricks workspace, select the Clusters and choose the Spark cluster created earlier.

3. Within that, select the Libraries and click Install New.

4. In the New library page, select source as Maven and enter the following coordinates:

   - Spark Event Hubs connector - com.microsoft.azure: azure-eventhubs-Spark_2.11:2.3.12

5. Click Install/

6. Create a Notebook in Spark named AnalyzeTweetsFromEventHub with Notebook type as Scala.

7. In the Notebook (Figure 6-14), tweets from Event Hub will be ready and cognitive services APIs will be called to perform sentiment analysis on the tweets.

Code:

```
//Cognitive service API connection String
    val accessKey = "<PROVIDE ACCESS KEY HERE>"
    val host = "https://cognitive-docs.cognitiveservices.azure.com/"
    val languagesPath = "/text/analytics/v2.1/languages"
    val sentimentPath = "/text/analytics/v2.1/sentiment"
    val languagesUrl = new URL(host+languagesPath)
    val sentimenUrl = new URL(host+sentimentPath)
    val g = new Gson

    def getConnection(path: URL): HttpsURLConnection = {
        val connection = path.openConnection().asInstanceOf[HttpsURLConnection]
        connection.setRequestMethod("POST")
        connection.setRequestProperty("Content-Type", "text/json")
        connection.setRequestProperty("Ocp-Apim-Subscription-Key", accessKey)
        connection.setDoOutput(true)
        return connection
    }
```

```
def prettify (json_text: String): String = {
    val parser = new JsonParser()
    val json = parser.parse(json_text).getAsJsonObject()
    val gson = new GsonBuilder().setPrettyPrinting().create()
    return gson.toJson(json)
}
```

8. The notebook will return output that will display text and
   sentiment value; the sentiment value will be in a range of 0 to 1.
   A value close to 1 suggests positive response, while one near to 0
   represents negative sentiments.

***Figure 6-14.*** *Create Event Hub instance*

The preceding exercise will provide hands-on experience with an Apache Spark
on Databricks cluster. Now, let's discuss the serverless offering called Azure Stream
Analytics on Microsoft Azure for real-time data processing.

# Stream Analytics

Stream analytics is a managed stream processing service natively built on Microsoft Azure. Stream analytics is one of the serverless offerings available on Azure. Since it's a serverless platform, there is no concept of CPU and memory. The unit of compute is streaming units, which is a combination of CPU and memory resources. Moreover, it supports declarative query language (i.e., SQL), which helps to extract output with just a few lines of code. To process the streaming data, a stream analytics job must be created. There are four major components of a stream analytics job:

1. Inputs

2. Functions

3. Query

4. Output

**Inputs** – Stream analytics can accept inputs from streaming platforms like Event Hubs and IoT Hubs, and from Blob storage as well. Event Hubs and IoT Hubs have been discussed in the earlier chapters. Let's discuss when Blob storage will be used as an input to stream an analytics engine. Blob storage is generally for keeping offline data; but for a log analytics scenario, data can be picked from Blob storage as data streams. This can help to process a large number of log files using stream analytics.

Another important data input is reference data. Reference data can be used to make joins in the query to generate results that are more refined and relevant to the use case. Reference data can be picked from Azure SQL DB or an SQL managed instance or Blob storage.

**Functions** – To extend the functionality into stream processing, stream analytics offers the use of the following functions:

1. JavaScript user-defined functions

2. JavaScript user-defined aggregates

3. Azure ML service

4. Azure ML Studio

These functions can be called in the query section to perform certain operations that are outside the standard SQL query language.

**Query** – The SQL query is written under this section, which processes the input data and generates the output. For real-time processing, there are a few common windowing scenarios like sliding window, tumbling window, or hopping window. There are geospatial scenarios like identifying distance between two points, geofencing, and fleet management, etc., which can be done natively with the help of functions available in a stream analytics job. Scenarios like transaction fraud or identifying any anomaly in the data can be done by invoking Azure ML service or ML Studio functions in the query.

**Output** – The stream processing output lands in the model and serve phase. There are three major scenarios that are built under stream processing:

1. Data storage for batch mode processing

2. Real-time dashboards and end-user notification

3. Automation to kick off workflow

    **Data storage for batch mode** – For this purpose, stream analytics output can be stored in Azure Blob storage, Azure Data Lake Storage, Azure SQL DB, MI, etc. After the data is stored in the storage layer, it can be further processed with Apache Spark on HDInsight, Databricks platform, or Synapse analytics platform, etc.

    **Real-time dashboards and end user notification** – Stream analytics output can also land directly to power BI, Tableau, etc. for real-time dashboards. For end-user notifications, this output can land in Cosmos DB. Scenarios like the stock market, where users have to be to buy or sell stocks, or currency rate change that happens in real time and has to be notified to the traders, can be managed efficiently.

    **Automation to kick-off workflow** – For automation to kick off workflow, Event Hubs can be used as an output or directly invoke Azure functions for performing operations like emailing a user or executing certain batch files, etc.

## Exercise: Sentiment Analysis on Streaming Data Using Stream Analytics

Let's see how a stream analytics job can be created to process the streaming data.

In Chapter 4, an exercise to do sentiment analytics for COVID-19 tweets was started. Phase-wise steps of the exercise were as follows:

1. *Data Ingestion*: Create Event Hub and ingest event data using an application

2. *Prep and Train*:

    a. Create stream analytics job and extract the Covid-19 related tweets

    b. Create R script for sentiment analytics using Azure ML studio

    c. Invoke R script in stream analytics job to extract the sentiments

3. *Model and Serve*: Show the data in Power BI and Cosmos DB

Data ingestion was done into Event Hubs using an application to extract Twitter events. In this chapter, the remaining steps to perform the sentiment analytics under Prep and Train will be performed.

Let's create a stream analytics job (Figure 6-15):

1. In the Azure portal, create a stream analytics job.

2. Name the job and specify subscription, resource group, and location.

3. Choose the Hosting environment as Cloud for deploying the service on Azure.

4. Select Create.

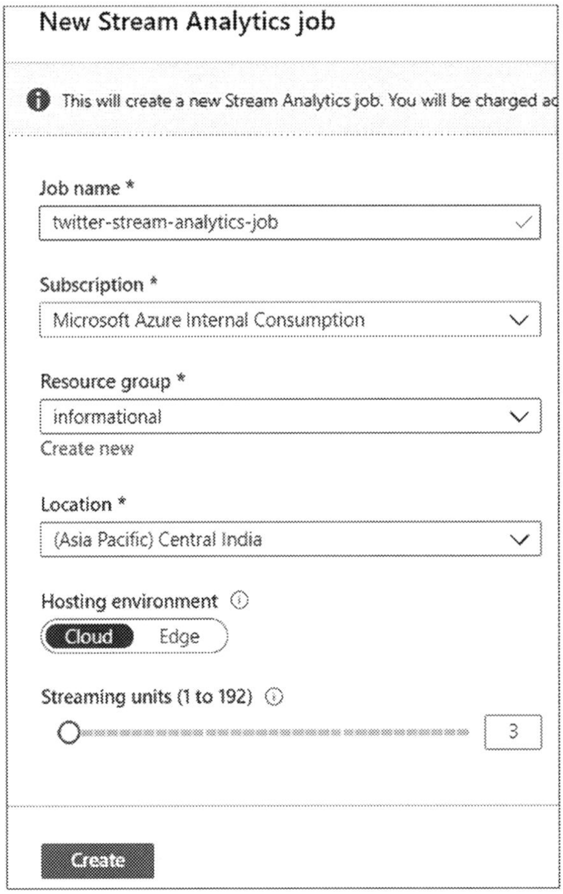

**New Stream Analytics job**

ⓘ This will create a new Stream Analytics job. You will be charged ad

Job name *

twitter-stream-analytics-job                          ✓

Subscription *

Microsoft Azure Internal Consumption                 ⌄

Resource group *

informational                                        ⌄
Create new

Location *

(Asia Pacific) Central India                          ⌄

Hosting environment  ⓘ

( **Cloud**   Edge )

Streaming units (1 to 192)  ⓘ

○━━━━━━━━━━━━━━━━━━━━━━━━━━━━  [ 3 ]

[ Create ]

***Figure 6-15.*** *Create stream analytics instance*

Once the stream analytics job service is created, the next step is to create a job (Figure 6-16):

1. Open the stream analytics; select Inputs from the menu.

2. Click Add Stream Input and fill in the required details like Input Alias, Subscription, namespace, name, policy name, and compression type.

3. Leave the remaining field as default and save.

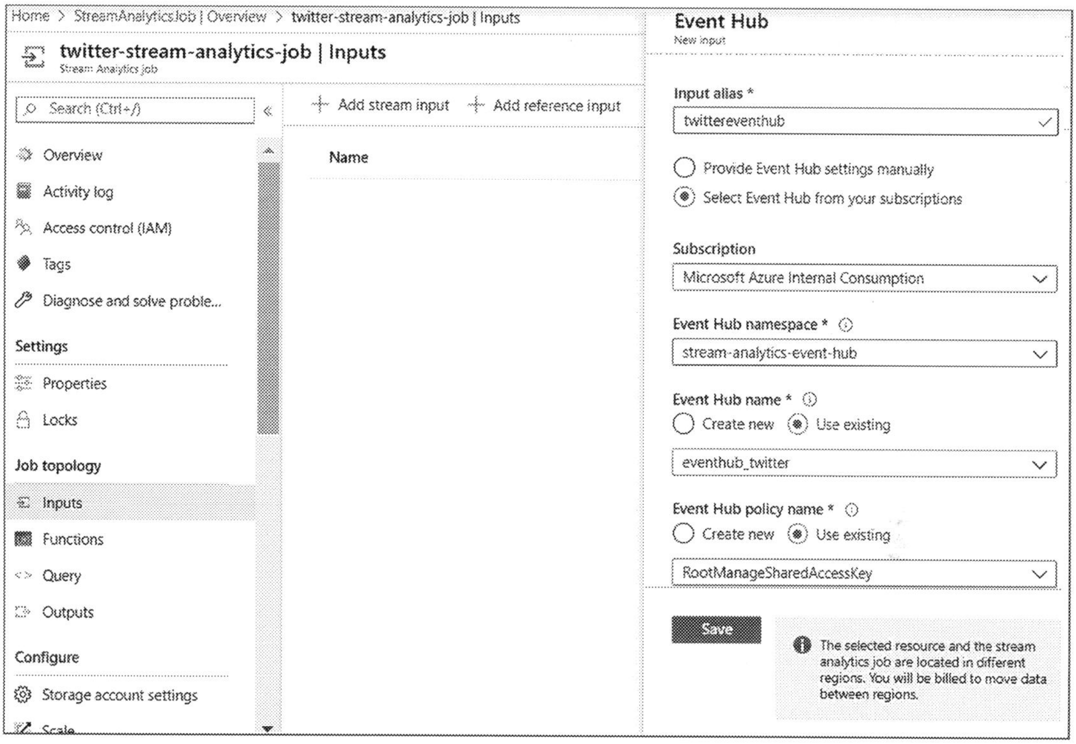

**Figure 6-16.** *Stream analytics–input*

Now you need to specify the job query. You can create simple, declarative queries for transforming or aggregating the data. For example, you can type queries like

```
Select * from twittereventhub;
```

This query returns all tweets that are there in Event Hub (Figure 6-17).

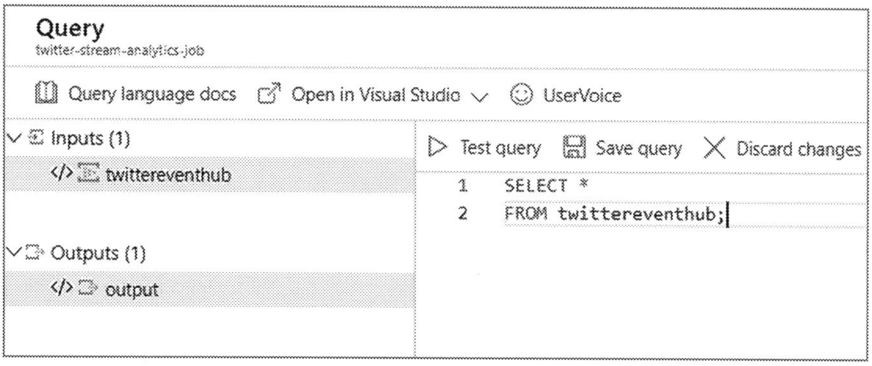

**Figure 6-17.** *Stream analytics–query*

```
SELECT System.Timestamp as Time, text
FROM twittereventhub
WHERE text LIKE '%COVID-19%';
```

This query returns all tweets that include the keyword COVID-19 (Figure 6-18).

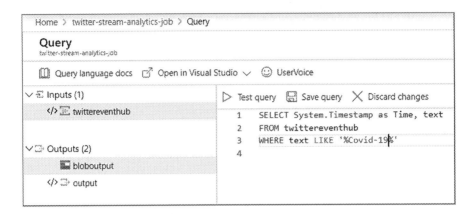

***Figure 6-18.***  *Stream analytics—query*

You can also push the results of the queries to an output sink, which could be
Azure Blob storage, Azure SQL Database, Azure Table storage, Event Hubs, or Power BI,
depending on your application needs (Figure 6-19).

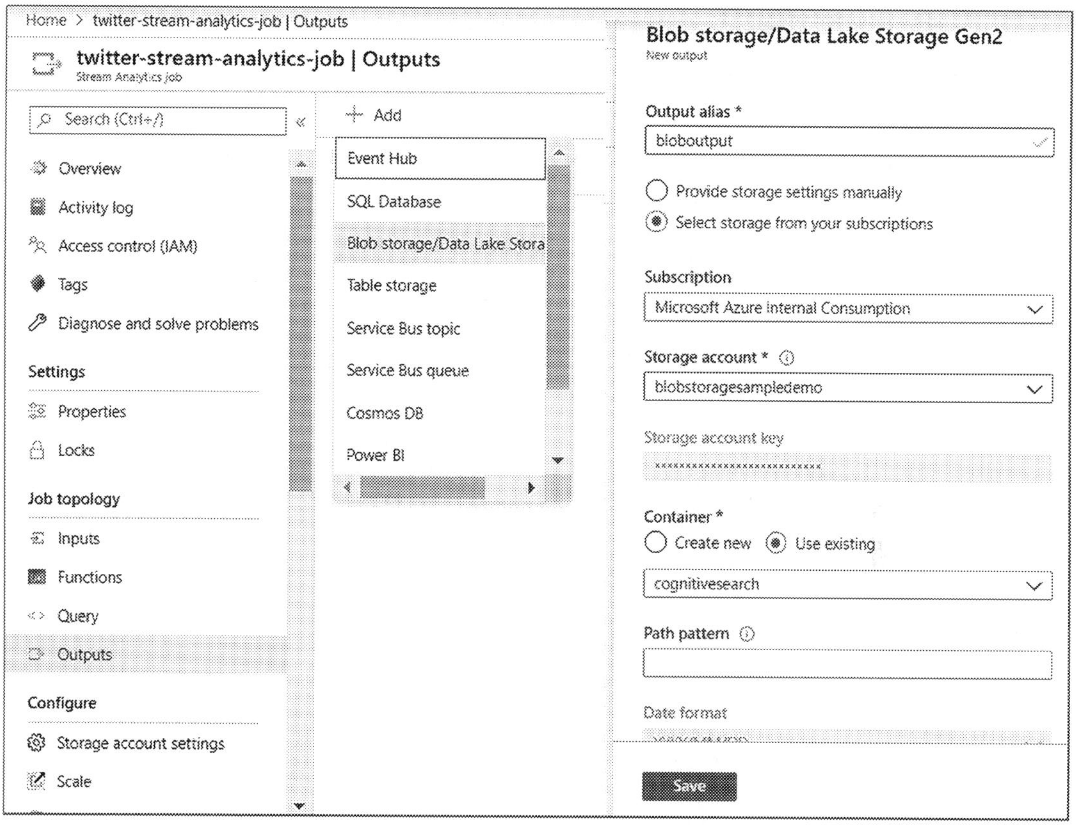

**Figure 6-19.** *Stream analytics—outputs*

Once the job is ready, you can start the stream analytics job and analyze the incoming tweets on a periodic basis.

Now you need to add the sentiment analytics model from the Cortana Intelligence Gallery.

1. Create ML workspace and go to Cortana Intelligence Gallery, choose the predictive sentiment analytics model and click Open in Studio.

2. Sign in to go to the workspace. Select a location.

3. At the bottom of the page, click the Run icon.

4. Once the process runs successfully, select Deploy Web Service.

5. Click the Test button to validate the sentiment analytics model. Provide text input such as "WHO has declared coronavirus as a pandemic."

6. The service will return a sentiment (positive, neutral, or negative) with a probability.

7. Now click the Excel 2010 or earlier workbook link to download an Excel workbook. This workbook contains the API key and the URL that are required later to create a function in Azure Stream Analytics.

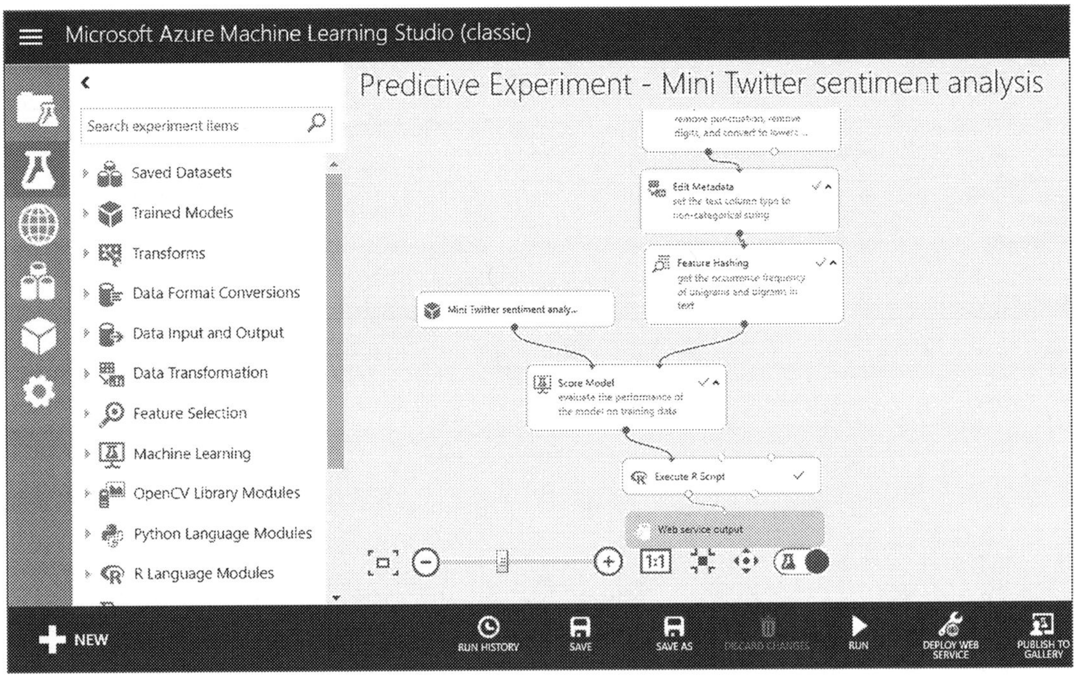

***Figure 6-20.*** *Azure Machine Learning Studio for predictive sentiment analytics*

Now go back to the stream analytics job to add a user-defined function (UDF; Figure 6-21). The function can be invoked to send a tweet to the web service and get the response back.

1. Go to your stream analytics job and choose Functions, then click Add and choose AzureML

2. Provide the required details like Functional Alias, URL, and Key. URL and Key are the same that you copied in the previous step.

3. Click Save.

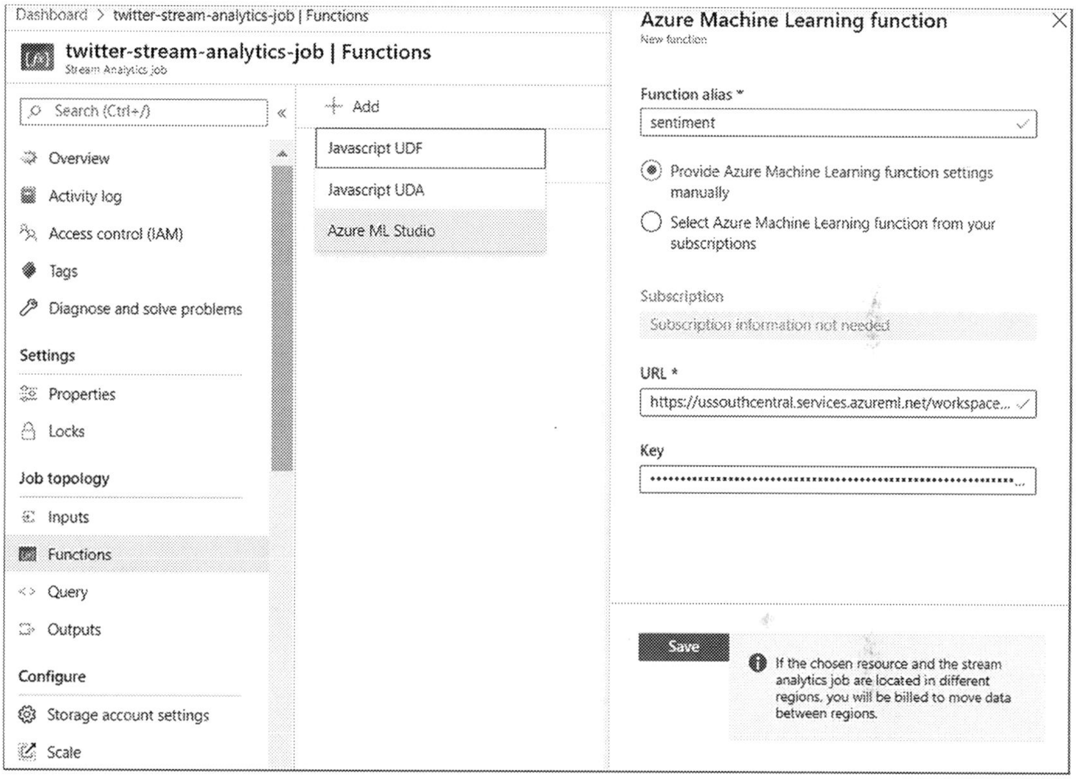

***Figure 6-21.*** *Azure Stream Analytics—user-defined function*

Now create another job query to include sentiment analysis.

1. Under Job Topology, click the Query box.

2. Enter the following query:

```
WITH sentimentdata AS (
SELECT text, sentiment(text) as result
FROM twittereventhub
)
```

```
SELECT text, result.[Score]
INTO bloboutput
FROM sentimentdata
```

3. The preceding query invokes the sentiment function so as to perform sentiment analysis on each tweet in the Event Hub.

4. Click Save query (Figure 6-22).

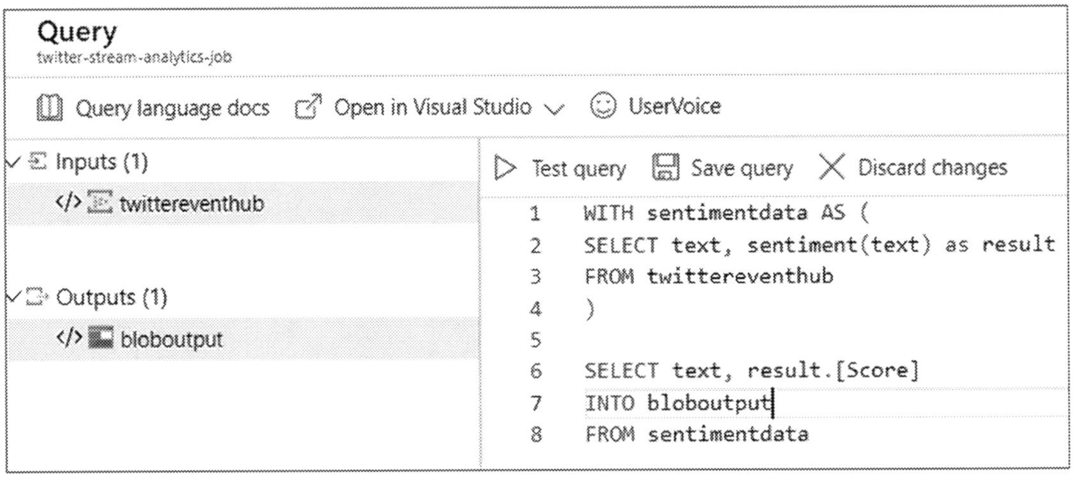

**Query**
twitter-stream-analytics-job

📖 Query language docs    ⬀ Open in Visual Studio ⌄    ☺ UserVoice

∨ ⬚ Inputs (1)               ▷ Test query  💾 Save query  ✕ Discard changes
    </> ⬚ twittereventhub
```
1    WITH sentimentdata AS (
2    SELECT text, sentiment(text) as result
3    FROM twittereventhub
4    )
5
6    SELECT text, result.[Score]
7    INTO bloboutput
8    FROM sentimentdata
```
∨ ⬚ Outputs (1)
    </> 🖼 bloboutput

***Figure 6-22.*** *Azure Stream Analytics—query*

The output looks like Figure 6-23.

| Text | Scored Probability |
|---|---|
| Current symptoms reported for patients with 2019-nCoV include acute onset of fever, cough, and difficulty in breathing. | ["neutral","0.466141670942307"] |
| There is currently no vaccine to prevent 2019-nCoV infection. The best way to prevent infection is to avoid being exposed to this virus. . All non-essential travel to China or affected countries* to be avoided. . Observe good personal hygiene. . Practice frequent hand washing with soap. . Cover your mouth when coughing and sneezing. *The list of affected countries is available on WHO website(www.who.int) and would be updated time to time. | ["negative","0.381420433521271"] |
| Self monitor your health starting from the day of last contact with such a case and continue for 28 days. Watch for the development of acute onset of signs & symptoms • Fever • Cough • Shortness of breath or difficulty in breathing If you observe any of the above symptoms visit the nearest health facility for further advice & treatment. Further you must furnish the details of exposure of such patient to your health care worker. | ["positive","0.936231553554535"] |

***Figure 6-23.*** *Azure Stream Analytics—query output*

This output can be projected directly to Power BI or stored in Azure SQL DB or any storage, as per the use case. This is further discussed in Chapter 8.

In summary, Azure Databricks and Azure Stream Analytics are widely used solutions to process real-time data. However, data prep concepts also call for data cleansing, data transformation, data reduction etc. In the preceding section, the discussion was around stream processing. For real-time processing, especially when the response must be real time, typical data preparation steps can't be afforded as it can induce latency. Only streaming data is processed or joined with multiple datasets or APIs, to cater to the target use cases. Data preparation techniques are followed during batch mode processing. There, the size of data is massive; data is of disparate types and all the data must be crunched together to derive a specific outcome. Let's discuss processing batch mode data in detail in the next section.

# Process Batch Mode Data

Batch mode data processing is a widely used scenario. Real-time data processing deals with millions of streams hitting the server in milliseconds, and the data processing must be done in real time. However, batch mode data processing deals with data coming from disparate applications, and the processing can be done at regular intervals or on-demand. In the recent past, lots of innovation has happened in this space. A decade ago, the majority of the solutions were based on monolithic applications built on top of structured data stores like SQL Server and Oracle, etc. Then, the only solution to process the batch data was by building enterprise data warehouses.

## Enterprise Data Warehouse

On conventional data warehousing systems, standard procedure to analyze large data was:

1. Build centralized operational data stores using ETL tools.

2. Understand KPIs and build data models.

3. Build a caching layer using OLAP platforms to precalculate KPIs.

4. Consume these KPIs to build dashboards.

In 2010, newer types of data stores built on NoSQL technologies were introduced. Then, analyzing large data of NoSQL technologies using Hadoop ecosystem used to be done.

Further to that, there was an evolution in application architectures from monolithic to SOA and now microservices, which promoted use of polyglot technologies. This is when the tipping point occurred; there was data of multiple types getting generated from a single application. Moreover, there were multiple independent LOB applications running in silos. Similarly, CRM applications and ERP applications generated lots of data, but no system was talking to each other.

With the help of big data technologies, this data was analyzed to an extent. Technologies like Apache Spark or other Hadoop-based solutions helped to analyze massive amount of data to produce meaningful output. The public cloud further accelerated the adoption of these solutions, and analyzing this large and disparate data became easier. This was called data lake analytics.

## Data Lake Analytics

A data lake is a combination of structured, semistructured, and unstructured data. Today, organizations want to build 360-degree scenarios where they want to integrate data from LOB applications, CRM, supply chain, and social media. Data is stored in CSV, Parquet, or JSON format on the storage layer. Then, using Spark or other Hadoop-based solutions, the data is transformed and then hosted in the storage layer for consumption by the downstream applications. The architecture of data lake analytics is shown in Figure 6-24.

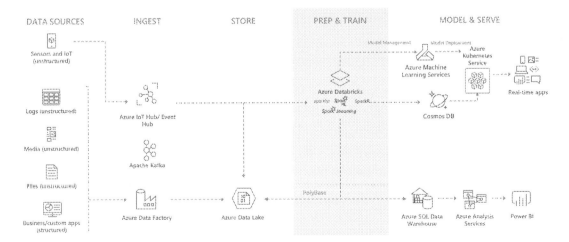

***Figure 6-24.*** *Data lake analytics*

The general rule of thumb is to analyze structured data with MPP architecture-based solutions like Azure Synapse analytics (Formerly known as Azure SQL DW) and unstructured and semistructured data with Hadoop-based distributed solutions like Apache Spark or Databricks., etc. However, Hadoop-based technologies can analyze structured data as well.

Let's discuss how the world of both enterprise data warehouse and data lake analytics is changing into modern data warehouse and advanced data lake analytics, respectively. Moreover, let's understand the key drivers that influence the decision to choose between the modern data warehouse and advanced data lake analytics.

## Modern Data Warehouse

Modern data warehouses deal with heterogeneous data coming from disparate data sources. Conceptually, data lake analytics or data analytics and enterprise data warehouses have merged into a single term called modern data warehouse. The gap between choosing MPP and distributed processing systems has been reduced to a great extent. Earlier, the choice was simple: if the data is structured, go for MPP platforms like Azure SQL DW or AWS Redshift, etc. If the data is unstructured and semistructured, go for Hadoop platforms like Spark or Cloudera data platform, etc.

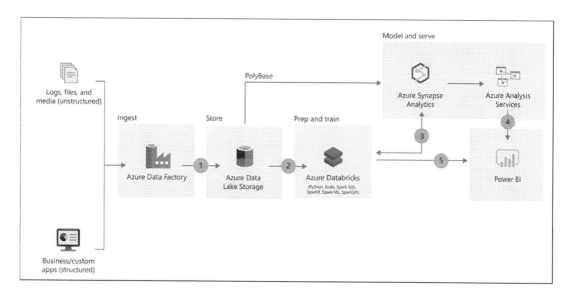

***Figure 6-25.*** *Modern data warehouse architecture*

As discussed earlier, with various types of applications working in silos, it's difficult to invest in an enterprise data warehouse running on MPP and then build another analytics platform running on Hadoop. Moreover, technical staff working on such technologies are expensive and scarce. Organizations prefer technologies that are cost effective, easier to learn, can bring data on a centralized platform, and they can leverage existing technical staff to manage these solutions. Therefore, it was necessary to build such platforms that can manage any kind of data. As shown in Figure 6-25, a modern data warehouse can manage any kind of data, and it can be a central solution to fulfill organization-wide needs.

Recently, Microsoft launched a technology solution on Azure called Synapse Analytics. This was formerly called, Azure Data Warehouse, which was built on MPP architecture. With Azure Synapse Analytics, not only structured but semistructured and unstructured data can also be analyzed. It supports both MPP SQL DW and distributed platform Apache Spark; and for dashboarding, Pipelines for ELT/ETL operations and Power BI can be easily integrated in the platform. Synapse analytics will be discussed in more detail in Chapter 8.

## Exercise: ELT Using Azure Data Factory and Azure Databricks

Let's discuss data preparation with the help of the following exercise to extract, load, and transform (ELT) data by using Azure Data Factory and Azure Databricks.

In this exercise we will learn about performing ELT operations using Azure Databricks. The data is read from a remote file system through Azure Data Factory, stored in Azure Data Lake Storage Gen2, and further processed in Azure Databricks.

The high-level architecture looks like Figure 6-26.

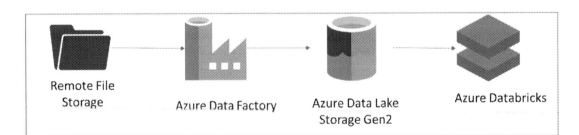

*Figure 6-26.* *Architecture for the ELT exercise*

The Azure services required for this exercise include:

1. Azure Data Factory

2. Azure Data Lake Storage (Gen2)

3. Azure Databricks

The exercise includes the following high-level steps:

1. Create an Azure Databricks workspace.

2. Create a Spark cluster in Azure Databricks.

3. Create an Azure Data Lake Storage Gen2 account.

4. Create an Azure Data Factory.

5. Create a Notebook in Azure Databricks to read data from Azure Data Lake Storage and perform ELT.

6. Create a pipeline in Azure Data Factory to copy data from a remote file system to Azure Data Lake Storage and then invoke Databricks notebook to perform ELT on the data.

Step 1 and step 2 to create an Azure Databricks workspace and Databricks cluster are the same as mentioned in the previous exercise.

For step 3—creating the Azure Data Lake Storage (Gen2)—the steps include:

1. In the Azure portal, create an Azure Databricks workspace by selecting **Create a resource ➤ Storage ➤ Storage Account**.

2. To create the Azure Databricks service, provide the required details (Figure 6-27), which are:

    • Subscription

    • Resource group

    • Storage account name

    • Location

    • Performance

    • Account kind

    • Replication

    • Access tier

133

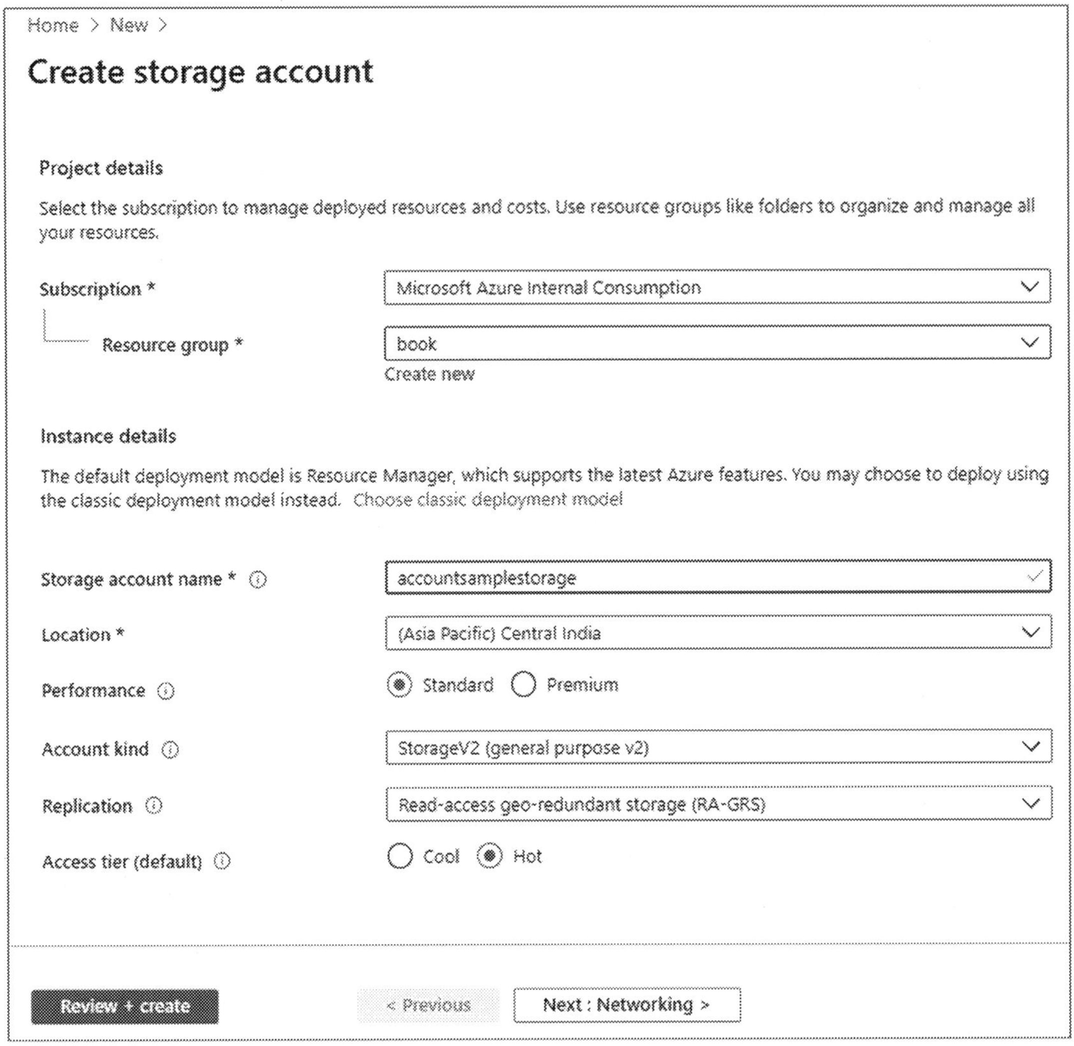

Home > New >

# Create storage account

### Project details

Select the subscription to manage deployed resources and costs. Use resource groups like folders to organize and manage all your resources.

Subscription *                          Microsoft Azure Internal Consumption                          ⌄

    Resource group *                book                                                          ⌄
                          Create new

### Instance details

The default deployment model is Resource Manager, which supports the latest Azure features. You may choose to deploy using the classic deployment model instead.  Choose classic deployment model

Storage account name *  ⓘ        accountsamplestorage                                         ✓

Location *                               (Asia Pacific) Central India                              ⌄

Performance  ⓘ                       ⦿ Standard  ◯ Premium

Account kind  ⓘ                       StorageV2 (general purpose v2)                          ⌄

Replication  ⓘ                        Read-access geo-redundant storage (RA-GRS)             ⌄

Access tier (default)  ⓘ             ◯ Cool  ⦿ Hot

[ Review + create ]        [ < Previous ]    [ Next : Networking > ]

***Figure 6-27.***  *Create storage account*

3.   Next is to enable the hierarchical namespace under the Advanced tab (Figure 6-28).

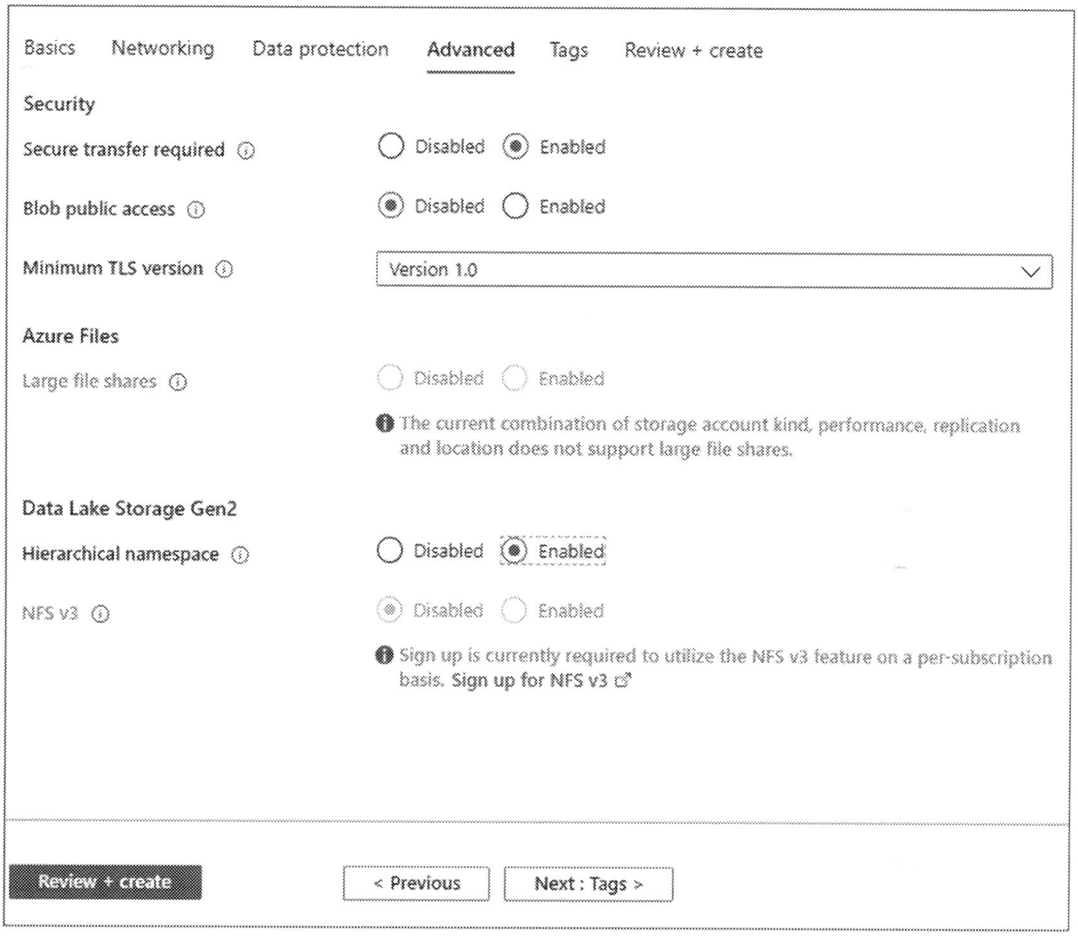

**Figure 6-28.** *Create storage account—advanced*

4.  Click the Review + create button to create the storage account.

Step 4 is to create an Azure Data Factory service (Figure 6-29) by doing the following steps:

1.  In the Azure portal, create an Azure Databricks workspace by selecting **Create a resource ➤ Analytics ➤ Data Factory**.

2.  To create the Azure Databricks service, provide the required details, which are:

    •   Name

    •   Version

- Subscription

- Resource group

- Location

- Enable Git

3. Press Create button to create

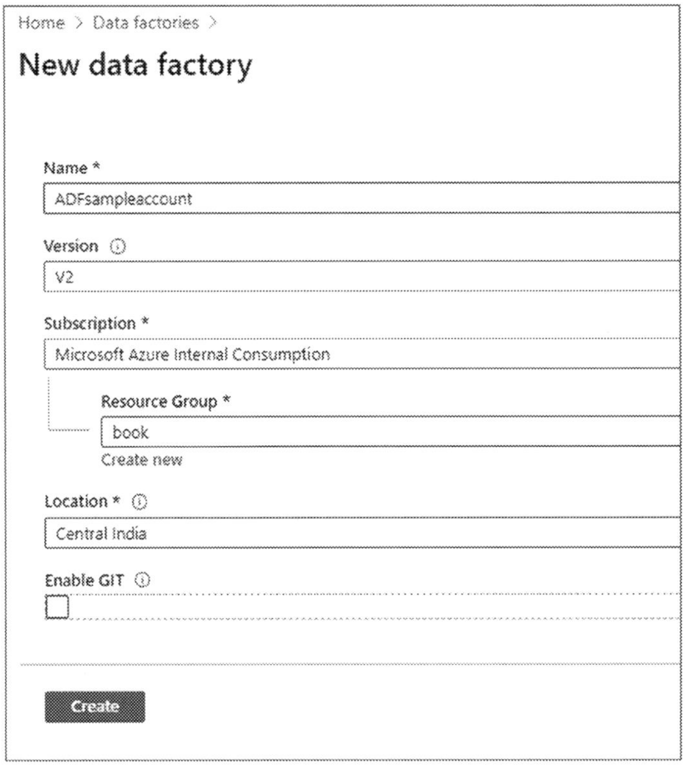

**Figure 6-29.** *Create data factory*

In step 5, you need to create a notebook in Azure Databricks, which can be used to perform ETL operations on the data:

1. Open the Azure Databricks workspace and create a Scala notebook.

2. The notebook is used to perform the following operations:

    a. Connect to ADLS Gen2 storage using the keys.

      b.  Read csv data from the Storage

      c.  Load CSV data into the Spark SQL delta table.

      d.  Perform insert into the delta table.

      e.  Perform aggregation on the data.

      f.  Save the aggregated data back to ADLS Gen2.

3.  The sample notebook is available at `https://github.com/`
    `Apress/data-lakes-analytics-on-ms-azure/blob/master/`
    `ETLDataLake/Notebooks/Perform%20ETL%20On%20ADLS(Gen2)%20`
    `DataSet.scala`

In step 6, let's create a pipeline in the Azure Data Factory service using the following steps:

1.  Go to the Azure Data Factory instance created in an earlier step and click Author and monitor.

2.  In the Data Factory UI, create a new pipeline.

3.  This pipeline is intended to perform the following operations:

      a.  Copy data from remote SFTP location to Azure Data Lake Storage Gen2.

      b.  Invoke the Databricks notebooks created earlier to perform ELT operations on the data.

4.  The process looks like Figure 6-30.

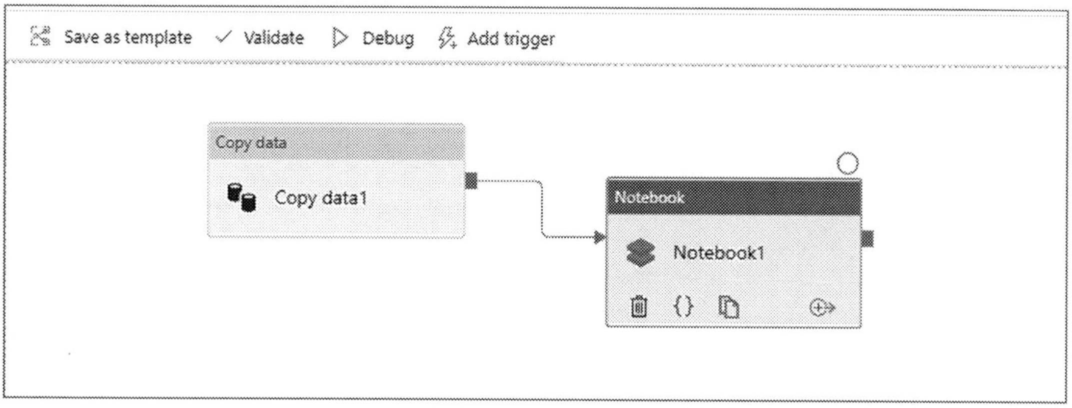

***Figure 6-30.*** *Create Azure Data Factory*

5.  To build the preceding pipeline, first you need to use a Copy Data activity, which can copy the data from source to sink: source being remote SFTP server and sink is Azure Data Lake Storage Gen2.

6.  Create an SFTP-linked service (Figure 6-31) by passing the following parameters:

    a.  Name

    b.  Host

    c.  Port

    d.  Username

    e.  Password

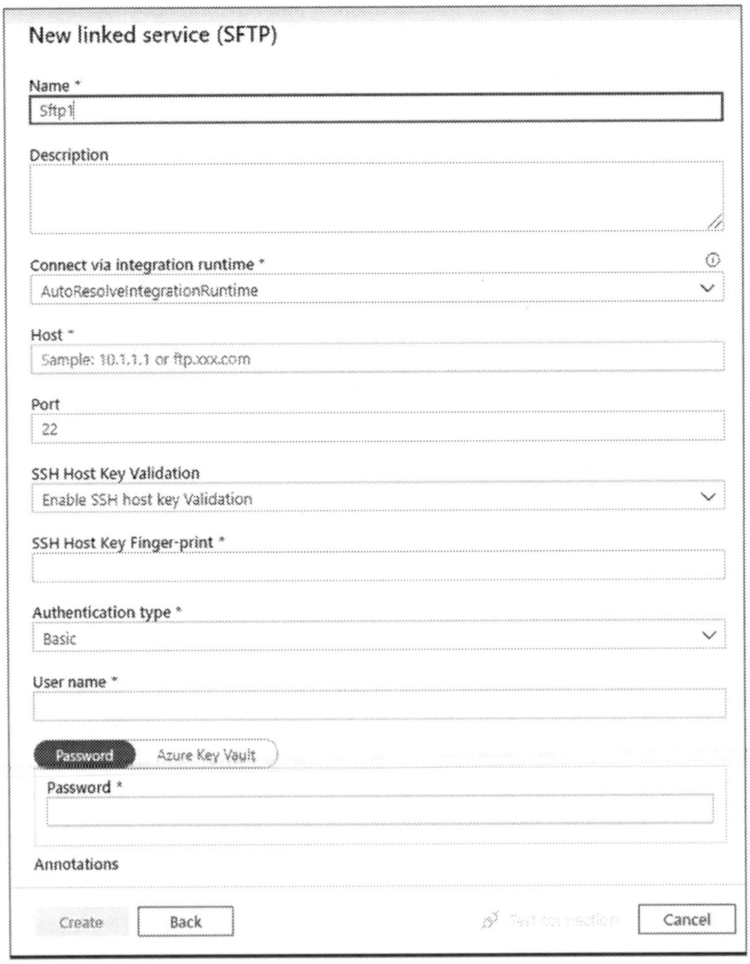

***Figure 6-31.*** *Create SFTP-linked service*

7. Create an Azure Data Lake Storage Gen2-linked service (Figure 6-32) by passing the following parameters:

   a. Name

   b. Azure subscription

   c. Storage account name

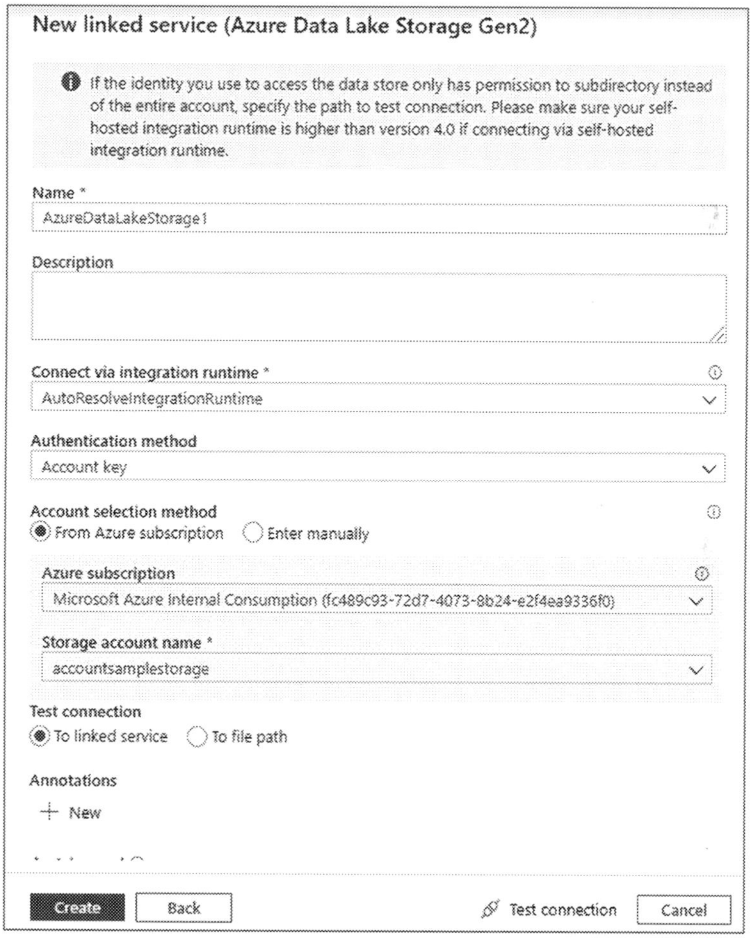

*Figure 6-32.* *Create Azure Data Factory-linked service*

8.  Create an Azure Databricks-linked service (Figure 6-33) by passing the following parameters:

    a.  Name

    b.  Azure subscription

    c.  Databricks workspace

    d.  Cluster type

    e.  Access token

    f.  Cluster ID

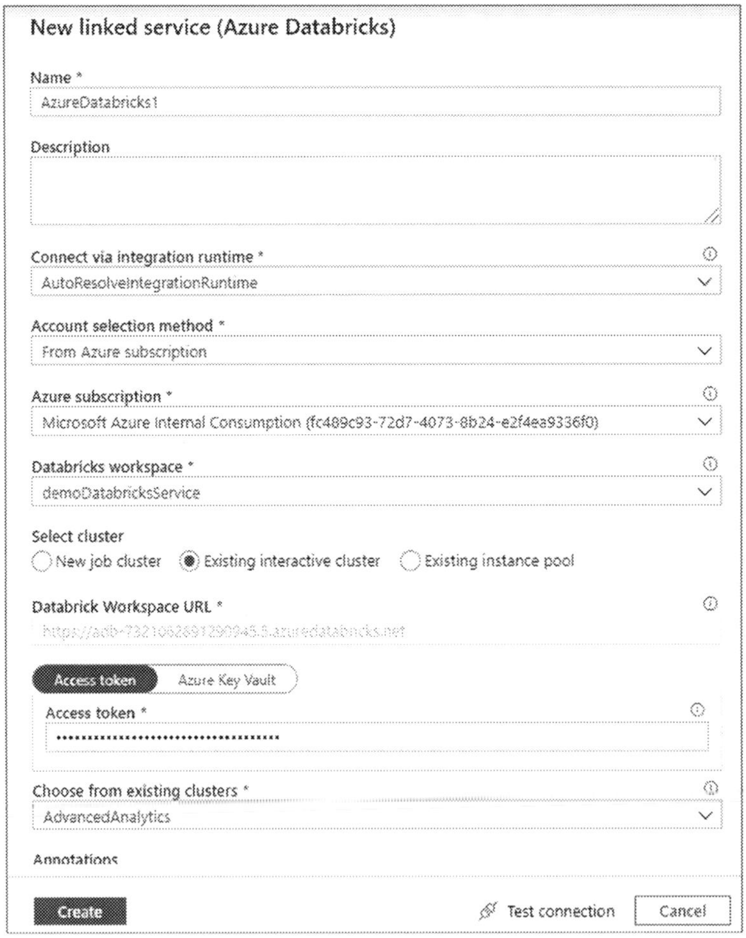

***Figure 6-33.*** *Create Azure DataBricks-linked service*

9. Configure the copy data activity by selecting the

    a. source dataset through the SFTP-linked service.

    b. sink dataset through the Azure Data Lake Storage Gen2-linked service (Figure 6-34).

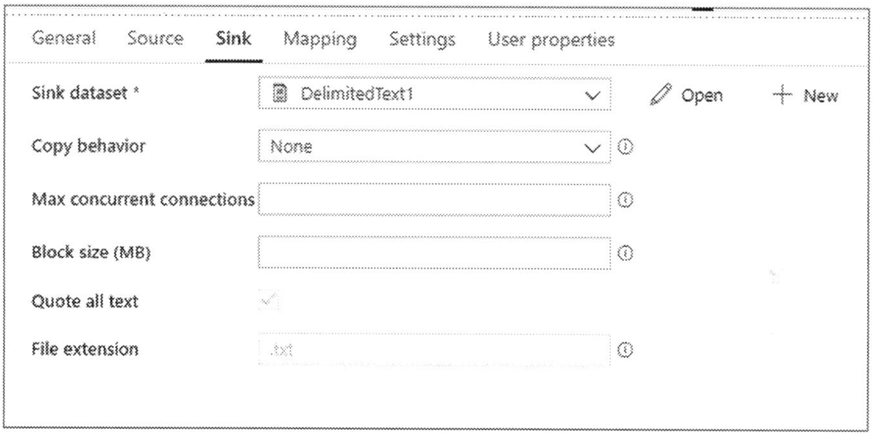

***Figure 6-34.*** *Provide the sink details*

10. Configure the Databricks notebook activity by

    a. selecting the Databricks-linked service created in the earlier step (Figure 6-35).

***Figure 6-35.*** *Select Azure Databricks-linked service*

    b. selecting the Databricks notebook created in earlier step for performing the ELT operation (Figure 6-36).

| General | Azure Databricks | **Settings** | User properties | | |
|---|---|---|---|---|---|
| Notebook path * | | /Users/pakhatta@microsoft.com/book/Perl | | 🗁 Browse | Open |
| ▷  Base parameters | | | | | |
| ▷  Append libraries | | | | | |

***Figure 6-36.***   *Select Azure Databricks notebook*

11.   Once the pipeline is created, save, publish, and trigger the
pipeline to monitor the results.

With this exercise, aggregated data is stored back on ADLS Gen2 storage, which
further can be picked up under the model and serve phase. It can be picked up by
Synapse Analytics using PolyBase, as shown in Figure 6-3, or directly picked up by power
BI for dashboarding. In the model and serve phase, there will be an exercise on picking
up this data in Synapse, caching in Analysis services, and showcasing in Power BI
dashboard.

# Summary

As discussed, this is the most critical phase in a data analytics solution. This phase
has got lots of new innovative technologies to build cost-effective and highly efficient
solutions. A decade ago, data warehouse and data mining of homogeneous data was
the only way to data analytics. The solution was effective but was expensive and time
consuming to set up. After moving to the cloud, the options to use MPP or distributed
systems became two major options. With the transformation in the developer space,
where the data was disparate, building a centralized solution was difficult. With
technologies like Apache Spark, Databricks, and Azure Synapse Analytics, this journey
has become easier and cost effective. Moreover, these technologies support multiple
programming languages, which makes the developer ecosystem large and accessible.
With the help of these solutions, nomenclature has now changed from data analytics and
enterprise data warehouse to modern data warehouse. The next chapter discusses what
advanced data analytics is, how the data can be prepared to train the machine learning
models, and the technologies available on Azure for building advanced analytics
solutions.

# CHAPTER 7

# Data Preparation and Training Part II

In the previous chapter, the discussion was on how the advent of new technologies such as IoT, streaming, or applications with disparate data structures brought lots of innovative technologies for data analytics. How the transformation from data analytics and enterprise data warehouse to modern data warehouse and advanced data analytics has happened. In part I of the prep and train phase, the discussion was on the modern data warehouse. In this chapter, let's delve deeper into advanced data analytics and the technologies available on Microsoft Azure that make building these solutions easier.

## Advanced Data Analytics

Data analytics solutions help us to understand the data and draw conclusion to make informed business decisions. It can help us to understand events that occurred in the past. However, to understand what can happen in the future with high probability or how to avoid specific events in the future is where advanced data analytics comes in. The advanced data analytics solution architecture remains the same as with data analytics; the only addition is machine learning (ML) technologies. When ML is infused in the data analytics pipeline to build prescriptions and predictions, the result is advanced analytics solution. In the prep and train phase, data is prepared to train the ML model. There are various platforms on Azure to build and deploy these models.

Let's discuss the basic concepts of data science and various platforms available on Azure to build ML models that can be infused into the data pipeline to build advanced analytics solutions. In Figure 7-1, the entire solution architecture is the same, with the addition of ML technologies like Spark ML.

© Harsh Chawla and Pankaj Khattar 2020
H. Chawla and P. Khattar, *Data Lake Analytics on Microsoft Azure*,
https://doi.org/10.1007/978-1-4842-6252-8_7

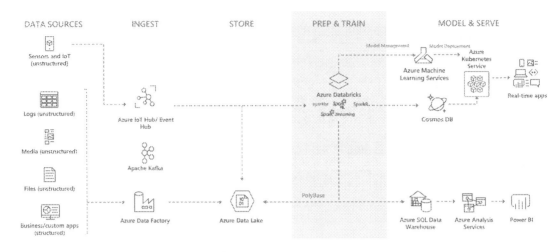

**Figure 7-1.** *Advanced data analytics*

In a way, data ingestion, data storage, data preparation, and data model and serve will remain the same. The only change is the addition of ML model training and consumption. Let's discuss the basics of data science and how it plays a vital role in building predictions and prescriptions.

# Basics of Data Science

Data science is the field that involves extracting information or insights from the large amount of available data by using algorithms and scientific methods. The idea is to discover patterns from the large amount of data without being explicitly programmed to do so.

Due to the emergence of big data, data analysis, and data statistics, the term "data science" became even more popular. The overall objective of using the data science process is to solve a business problem using ML algorithms, with a focus on predicting future events and forecasting trends.

Data science is becoming an essential field, as companies produce larger and more diverse datasets. For most enterprises, the data discovery process begins with data scientists diving through massive sets while seeking strategies to focus them and provide better insights for analysis.

There are multiple use cases across domains and industries that utilize data science to resolve business domain problems:

1.  *Travel*: Dynamic pricing, predicting flight delay

2. *Marketing*: Predicting lifetime value of a customer, cross-selling, upselling, customer churn, sentiment analysis, digital marketing

3. *Healthcare*: Disease prediction, medication effectiveness

4. *Sales*: Discount offering, demand forecasting

5. *Automation*: Self-driving cars, drones

6. *Credit and insurance*: Claims prediction, fraud, and risk detection

Let's delve deeper into the data science concepts in this section. The focus will be on the following:

1. Data science process

2. Personas in the data science process

3. Machine learning overview

**Data science process** – Data science is a cyclic process that consists of the following steps:

1. Define goal

2. Acquire and understand data

3. Preprocessing

4. Missing values

5. Exploratory data analysis

6. Feature engineering

7. Model building

8. Model evaluation

9. Model deployment

**1. Define the goal**

The first step in the process is to understand the business problem from the stakeholders and define an objective. Most of the time, the inputs are ambiguous and it's important to have a clear objective or problem statement. Another important aspect is to identify key stakeholders associated with the problem being solved. At the end of this step, all the information and context that are essential to solve the problem should be available.

## 2. Acquire and understand the data

This step involves discovering and accessing the required data from all the internal and external sources. This data can be in the form of logs, social media feeds, structured datasets, or any type of data source. This is the most important step in this entire process. Having the right set of data to solve the business challenge is half the job done.

## 3. Preprocessing the data

After the raw data is available, it's important to process that data. The raw data will have lots of inconsistencies like missing values, blank columns, and incorrect data format that need to be rectified. It's important to process the data for better outcome. As discussed in the previous chapter, the following steps need to be followed to prepare the data to train ML models.

## 4. Missing values

It's important that all the values are complete in the data, so dealing with missing values is crucial to building a model. The simplest solution is to remove all the rows that have a missing value, but that means that important information could be lost. Alternatively, you can impute a value (i.e., substitute a value for a missing value); the substitute value could be a mean, median, or a mode.

1. *Data cleaning*: Process to detect and remove noisy, inaccurate, and unwanted data

2. *Data transformation*: Data is converted to a format an ML model can read, which can involve changing the values to a standardized format. Data transformation techniques include categorical encoding, dealing with skewed data, scaling, or bias mitigation.

3. *Data reduction*: Data reduction is aggregation of data and identifying data samples needed to train the ML models.

4. *Data discretization*: Process to convert data into right-sized partitions or internals to bring uniformity

5. *Text cleaning*: Process to identify the data based on the target use case, and further cleaning data that is not needed.

## 5. Exploratory data analysis (EDA)

EDA can be an extremely useful step. Often, this part of the cycle will give insight into what needs to be done during feature engineering and modeling to produce the best results. You can use different techniques, depending on the types of data you are working with. A few common visualization techniques are histograms, distribution plots, box plots, and heat maps.

EDA helps to define the target variables. Discuss whether the target variable is to be transformed according to the objective. What is the distribution of the target variable? Is it binary or a continuous value? Depending upon these, the relationship between input variables is determined. This crucial step of EDA allows for the investigation of what is happening in the data that is not obvious. Sometimes it's possible to uncover patterns within the data, build questions about the data, and reject or accept the original hypotheses about the data.

## 6. Feature engineering

Feature engineering is a large part of the modeling procedure. Creating features using the data available often goes hand in hand with EDA because when you create a feature, it's important to see how it relates to the rest of the data. Therefore, you may find yourself going back and forth between EDA and feature engineering.

Some techniques used in feature engineering are moving averages and different types of aggregations. A moving average is the change in average for a specific constant time interval. Aggregations are combinations of the data based on another feature. Some examples of aggregations arc sum, average, and count.

A feature is an attribute or property of data that can be helpful to improve the performance of an ML model. Feature engineering helps to extract patterns from the data that were not visible earlier and identify new features that can help to improve accuracy of the ML model.

### 7. Model building

In this step, the actual model building starts. Based on the problem being solved, a relevant machine learning algorithm needs to be applied. Algorithms can be built based on whether numerical or classification prediction is being done. Next, you split the data into training and testing data sets. While training data is used to train the model, the testing data is used to evaluate the model for performance and accuracy.

Cross-validation is another common practice to improve the model accuracy and performance. Data is split into subsets, to ensure that the model is not overfitted to one training set.

### 8. Model evaluation

Once the model is built, model evaluation helps to understand the accuracy of prediction based on performance parameters of algorithms. A few examples of performance parameters are shown in Table 7-1.

***Table 7-1.*** *Performance Parameters*

| Classification | Regression | Time Series Forecasting |
| --- | --- | --- |
| accuracy | spearman_correlation | spearman_correlation |
| AUC weighted | normalized_root_mean_squared_error | normalized_root_mean_squared_error |
| average_precision_score_weighted | r2_score | r2_score |
| norm_macro_recall | normalized_mean_absolute_error | normalized_mean_absolute_error |
| precision_score_weighted | | |

Given that model training and validation is an iterative process, to arrive at desirable accuracy and precision, one needs to iteratively tune or change the parameters related to the algorithm. These parameters related to algorithms are hyperparameters. Hyperparameters are the parameters of the algorithm that cannot be learned but must be passed to the algorithm.

Hyperparameters control how the model training is done, which has a direct impact on model accuracy.

During model evaluation, the major focus is toward measuring the accuracy and precision of the model. While accuracy is the value of accurate predictions over the total predictions, precision is a measure of the consistency of the predictions.

Note: It's worth mentioning here that building an ML model with desirable accuracy and precision is often a cycle of data understanding, feature engineering, modeling, and model evaluation. Each of the aforementioned steps in the modeling cycle can take a significant amount of time (Figure 7-2).

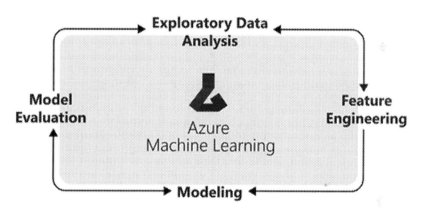

***Figure 7-2.*** *ML model building cycle*

### 9. Model deployment

This is the final step in the data science process. It typically involves deploying the final model into an environment or platform where it can be consumed by the applications to predict results for real-time or future data.

There are many considerations for deploying the model, depending upon the scale and frequency of its consumption. It could be as simple as deploying it as a web service or as complex as integrating it in a live web application or big data pipelines.

There is also a concept of MLOps, which is related to model deployment. MLOps, or DevOps for ML, enables data scientists and developer teams to collaborate and increase the pace of model development and deployment via monitoring, validation, and governance of ML models. MLOps provides the following capabilities:

- It helps in training reproducibility with advanced tracking of datasets, code, experiments, and environments.

- Just like DevOps, it supports deployments with continuous integration/continuous deployment (CI/CD) and also provides efficient workflows with scheduling and management capabilities.

- Advanced capabilities to meet governance and control objectives are also included.

**Data science personas** – There are many specialized personas in the data science process, usually consisting of a team of people with multiple roles:

1. *Business analyst*: A BA is the domain expert who provides business understanding and guides the overall project. They understand the functional areas and the data required to train an ML model.

2. *Data engineer*: A data engineer is the one who prepares the data required to train the model. All the major steps involved regarding data discovery, data access, data wrangling and cleansing are performed by the data engineer. They are expert in ETL operations and big data technologies.

3. *Data scientist*: A data scientist is the one who performs the EDA and data mining on the available data and decides which ML algorithm is chosen to create the model. They understand the overall data science process of building and tuning an ML model for the desired outcome.

4. *MLOps engineer*: MLOps engineers are responsible for deploying and consuming the model on a platform, based on the use cases (e.g., if the model has to be used in a website to detect real-time frauds or if the model has to be used in a big data pipeline to build predictions).

# Machine Learning and Deep Learning Overview

Machine learning is the process of building algorithms that improve automatically through experience by feeding them with structured data and manually performing feature extraction. There are multiple ML methods and algorithms that can be used to address the different types of problems.

Some of the examples of ML include image recognition, recommendation systems, speech recognition, etc.

Deep learning is a subset of ML where algorithms are created and function similarly to ML; but there are many levels of these algorithms, each providing a different interpretation of the data it conveys. This network of algorithms is called an artificial neural network (ANN). In simple words, it resembles the neural connections that exist in the human brain. In deep learning, the feature extraction is performed by the neural network itself.

An example of deep learning is that of driverless cars, which can detect objects on the road while driving, traffic signals, stop signs, pedestrians, etc. (Figure 7-3).

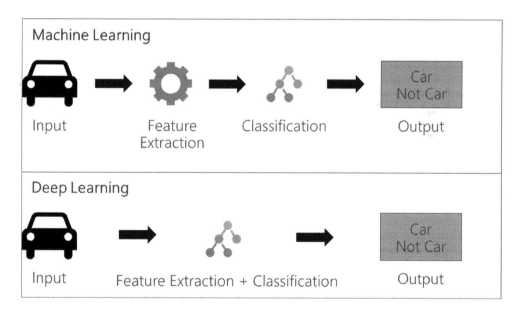

***Figure 7-3.*** *Machine learning vs. deep learning*

Let's also talk about types of machine learning:

- *Supervised learning*: It is when the model learns from a labeled dataset with guidance. Labeled dataset means that for each piece of data there is a label provided that tells the nature of the data; this helps the model in learning and hence predicting the solution easily. For example, a labeled data set could be of animal images, which trains the model to identify the type of animal. Whenever a new data image comes in, the trained model can compare its features with the labeled dataset to predict the correct label. There are two types of problems that supervised learning deals with; they are classification and regression:

  - *Classification*: The algorithm predicts a discrete value that can identify the input data as part of a group. For example, classifying that an image of an animal is of a cat or dog.

  - *Regression*: The algorithm is meant for continuous data—for example, predicting the price of a car, stock market value of a share, etc.

- *Unsupervised learning*: In unsupervised learning, data is not labeled. It's a type of self-organized learning that identifies unknown patterns in the data without preexisting labels. Grouping of data is done, and comparison is made by the model to guess the output. For example, to identify the animal as cat or dog, instead of labeled data, some feature information is used—like if an animal has features like floppy ears, a curly tail, etc., then it's a dog; if the animal has a flexible body, sharp teeth, etc., it's a cat.

- *Reinforcement learning*: It's a type of learning that is based on interaction with the environment. It is rapidly growing, along with producing a huge variety of learning algorithms that can be used for various applications. To begin with, there is always a start and an end state for an agent (the AI-driven system); however, there might be different paths for reaching the end state, like a maze. This is the scenario wherein reinforcement learning is able to find a solution for a problem. Examples of reinforcement learning include self-navigating vacuum cleaners, driverless cars, scheduling of elevators, etc.

The major objectives and main points of reinforcement learning are:

- Taking suitable action to maximize the rewards

- Learning from the experience

- One example is building models for the game of chess.

- Input: initial state from which the model starts

- Output: There are many possible outputs.

- Training: Based upon the input, the model returns a state and user reward.

- It's a continuous learning process.

- Best solution is based on the maximum rewards.

- Types of reinforced learning are positive and negative reinforcement.

A common task that most data scientists have is which algorithm to choose. To explain in simple terms, it primarily depends upon two aspects:

1. What is the business use case that you are trying to solve? And which category of ML (supervised, unsupervised, reinforcement, classification, regression, etc.) does it fall into?

2. What are the requirements to be achieved? What is the accuracy, training time, linearity, number of parameters, and features available?

Some of the common challenges that data scientists face in order to achieve the above tasks include:

- Lack of relevant data for the use case

- Quality of the available data

- A laborious process of hyperparameter tuning

- Model explainability, where the linear model explainability is easy but complex models providing highest accuracy are hard to interpret

- Following design principles for a real-time ML framework

Just for information, some of the algorithms are mentioned in Table 7-2.

***Table 7-2.*** *Algorithm Family*

| Algorithm | Accuracy | Training time |
|---|---|---|
| **Classification family** | | |
| Two-class logistic regression | Good | Fast |
| Two-class decision forest | Excellent | Moderate |
| Two-class boosted decision tree | Excellent | Moderate |
| Two-class neural network | Good | Moderate |
| Two-class averaged perceptron | Good | Moderate |
| Two-class support vector machine | Good | Fast |
| Multiclass logistic regression | Good | Fast |
| Multiclass decision forest | Excellent | Moderate |
| Multiclass boosted decision tree | Excellent | Moderate |
| Multiclass neural network | Good | Moderate |
| One-vs.-all multiclass | - | - |
| **Regression family** | | |
| Linear regression | Good | Fast |
| Decision forest regression | Excellent | Moderate |
| Boosted decision tree regression | Excellent | Moderate |
| Neural network regression | Good | Moderate |
| **Clustering family** | | |
| K-means clustering | Excellent | Moderate |

By now, data science concepts have been briefly covered. Now, let's understand various platforms available on Microsoft Azure to build ML models. There are four major managed (PaaS) services available on Azure:

1. Azure Machine Learning service:

    a. Azure Notebooks

    b. Azure Machine Learning designer service

    c. Azure Automated Machine Learning (AutoML)

2. Azure Databricks Spark MLlib

**Azure Machine Learning service**

Azure Machine Learning service is a fully managed cloud service that helps to easily build, deploy, and share predictive analytics solutions. It provides visual and collaborative tools to create predictive models that can be published as ready-to-consume web services, without worrying about the hardware or the VMs that perform those calculations.

Azure Machine Learning can be used to build classic ML models and deep learning for supervised and unsupervised learning. Depending upon the skillset and programming language knowledge of the data scientist, they can case choose from Python, R, or Scala to create their models. Data scientists can also use designer visual interface, which is no-code/low-code option to build, train, and track ML models. Many popular open source frameworks like PyTorch, TensorFlow, scikit-learn, etc. can be integrated and used in this service.

There are multiple available services in Azure for ML for model building, training, and deployment. As shown in Figure 7-4, there is a simple logical flow that suggests an appropriate service depending upon the availability of skill sets, resources, and time.

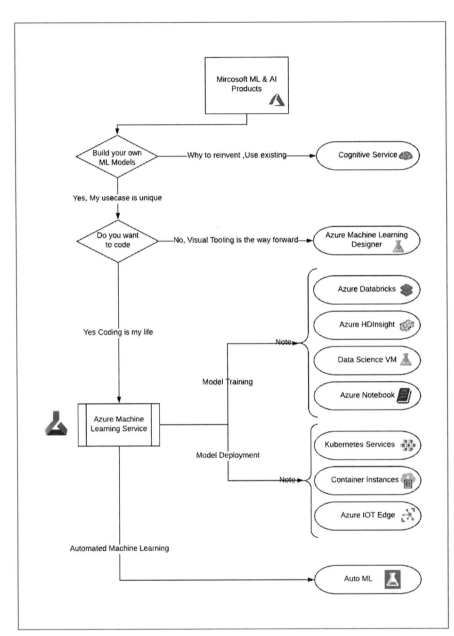

***Figure 7-4.*** *Microsoft ML and AI platform flowchart*

The service mentioned previously includes:

1. *Cognitive service*: Azure Cognitive Services are APIs, SDKs, and services available to help developers build intelligent applications without having direct AI or data science skills or knowledge. Azure Cognitive Services enable developers to easily add cognitive features into their applications. The goal of Azure Cognitive Services is to help developers create applications that can see, hear, speak, understand, and even begin to reason. The catalog of services within Azure Cognitive Services can be categorized into five main pillars: vision, speech, language, web search, and decision, and approximately 31 different types of APIs available.

2. *Azure Machine Learning Designer/Studio*: Microsoft Azure Machine Learning Studio is a collaborative, drag-and-drop tool you can use to build, test, and deploy predictive analytics solutions on your data. Machine Learning Studio publishes models as web services that can easily be consumed by custom apps or BI tools such as Excel.

3. *Azure Databricks*: Azure Databricks is an Apache Spark-based analytics platform optimized for the Microsoft Azure cloud services platform. Designed with the founders of Apache Spark, Databricks is integrated with Azure to provide one-click setup, streamlined workflows, and an interactive workspace that enables collaboration between data scientists, data engineers, and business analysts.

4. *Azure HDInsight*: Azure HDInsight is a managed, full-spectrum, open source analytics service in the cloud for enterprises. One can use open source frameworks such as Hadoop, Apache Spark, Apache Hive, LLAP, Apache Kafka, Apache Storm, R, and more.

5. *Data Science Virtual Machines*: The Data Science Virtual Machine (DSVM) is a customized VM image on the Azure cloud platform built specifically for doing data science. It has many popular data science tools preinstalled and preconfigured to jumpstart building intelligent applications for advanced analytics.

6.  *Notebook VM Web Service*: Azure Notebooks is a free hosted service to develop and run Jupyter Notebook in the cloud with no installation. Jupyter (formerly IPython) is an open source project that lets you easily combine markdown text, executable code, persistent data, graphics, and visualizations onto a single, sharable canvas—the notebook.

7.  *Azure Kubernetes Services*: Azure Kubernetes Service (AKS) makes it simple to deploy a managed Kubernetes clusters in Azure. AKS reduces the complexity and operational overhead of managing Kubernetes by offloading much of that responsibility to Azure. As a hosted Kubernetes service, Azure handles critical tasks like health monitoring and maintenance for you. The Kubernetes masters are managed by Azure. You only manage and maintain the agent nodes. As a managed Kubernetes service, AKS is free— you only pay for the agent nodes within your clusters, not for the masters.

8.  *Azure Container Instances*: Containers are becoming the preferred way to package, deploy, and manage cloud applications. Azure Container Instances offers the fastest and simplest way to run a container in Azure, without having to manage any virtual machines and without having to adopt a higher-level service.

9.  *Azure IoT Edge*: Azure IoT Edge is an IoT service that builds on top of IoT Hub. This service is meant for customers who want to analyze data on devices, or "at the edge," instead of in the cloud. By moving parts of your workload to the edge, your devices can spend less time sending messages to the cloud and react more quickly to events.

10. *Automated ML*: With automated machine learning, you can automate away time-intensive tasks. Automated ML rapidly iterates over many combinations of algorithms and hyperparameters to help you find the best model based on a success metric of your choosing.

**Azure Machine Learning Designer Service**

Azure ML Designer is a visual interface canvas where data scientists can connect data sets and modules to create ML models. The designer helps in performing the following operations through drag and drop:

*Pipelines*: Pipeline is used to train a single model or multiple models; it consists of datasets and analytical ML models that can be connected to each other. Pipelines can be reused and invoked in other pipelines.

*Datasets*: There are multiple datasets that are automatically included in the designer; the user can also upload their own datasets as well for model creation.

*Module*: They are the algorithms that can be applied on the datasets; there are multiple modules available, from data ingress functions to training, scoring, and validation processes. Each module has a specific set of parameters that can be used to configure the related algorithm.

*Compute resources*: They are required to execute your pipelines and host the deployed models for real-time or batch processing. The supported compute targets include:

- *Training*: Azure Machine Learning compute

- Azure Machine Learning compute instance

- *Deployment*: Azure Kubernetes Service

*Real-time endpoints*: Once created and completed, Pipelines can be deployed on Compute resources to get a real-time endpoint, through which external applications can call the model and get the predicted result or score.

*Published pipelines*: Further pipelines can also be published to a pipeline endpoint; they can be used to train and retrain a model through a REST call.

*Registered models*: Azure Machine Learning workspace keeps track of models through model registry. This is very helpful in maintaining the versioning, metadata, tags, etc.

The overall workflow is shown in Figure 7-5.

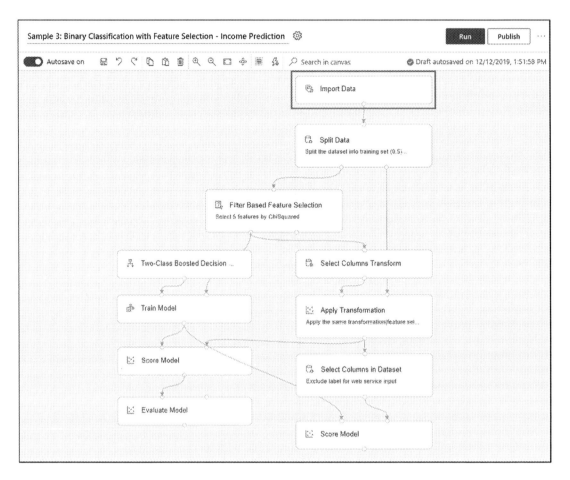

***Figure 7-5.*** *ML designer workflow*

Azure ML designer service provides a vast library of algorithm of classification, recommender system, cluster, anomaly detection, and regression to text analytics. There is an ML Algorithm cheat sheet that gives really good information to choose the right algorithms based on a target use case.

**Azure Automated Machine Learning**

As shared earlier, building an ML model is a cyclic and time-consuming task. AutoML helps in automating this overall development process. It helps in building ML models with high scale and efficiency, and in less time. AutoML also ensures that the quality of the model is upgraded.

To achieve high performance, modern ML techniques require careful data preprocessing and hyperparameter tuning. Moreover, given the ever-increasing number of ML models being developed, model selection is becoming increasingly important. Automating the selection and tuning of ML pipelines consisting of data preprocessing methods and ML models has been one of the goals of the ML community.

Automated ML can be used:

1. To create an ML model without extensive knowledge of all the available algorithms

2. To optimize Compute resources and time.

3. As an Agile way of creating an ML model

Azure Automated ML is offered in three categories

- *Classification*: Examples – fraud detection, marketing prediction

- *Regression*: Example – performance prediction

- *Time series forecasting*: Examples – sales forecasting, demand forecasting

Azure Automated ML experiments can be designed using the following steps:

1. Identify the problem to be solved: classification or regression, etc.

2. Upload the dataset and related metadata.

3. Configure the compute cluster by selecting the environment, which could be local computer, Azure VM, or Azure Databricks.

4. Configure the AutoML parameters, which include iterations over the models, hyperparameter settings, advanced featurization, exit criteria, and metrics to determine the best model.

5. Submit the experiment.

6. Review the results.

Moreover, feature engineering steps like feature normalization, handling missing data, converting text to numeric, etc. is applied automatically in AutoML but also can be customized based on the data.

Automated ML supports ensemble models, which are enabled by default. Ensemble learning improves ML results and predictive performance by combining multiple models, as compared with using single models. The ensemble iterations appear as the final iterations of the run.

Supported algorithms in Azure Automated ML are shown in Table 7-3.

***Table 7-3.*** *Supported Algorithms in Azure Automated ML*

| Classification | Regression | Time Series Forecasting |
|---|---|---|
| Logistic Regression | Elastic Net | Elastic Net |
| Light GBM | Light GBM | Light GBM |
| Gradient Boosting | Gradient Boosting | Gradient Boosting |
| Decision Tree | Decision Tree | Decision Tree |
| K Nearest Neighbors | K Nearest Neighbors | K Nearest Neighbors |
| Linear SVC | LARS Lasso | LARS Lasso |
| Support Vector Classification (SVC) | Stochastic Gradient Descent (SGD) | Stochastic Gradient Descent (SGD) |
| Random Forest | Random Forest | Random Forest |
| Extremely Randomized Trees | Extremely Randomized Trees | Extremely Randomized Trees |
| Xgboost | Xgboost | Xgboost |
| Averaged Perceptron Classifier | Online Gradient Descent Regressor | Auto-ARIMA |
| Naive Bayes | | Prophet |
| Stochastic Gradient Descent (SGD) | | ForecastTCN |
| Linear SVM Classifier | | |

### Azure Databricks Spark MLlib

Azure Databricks is a Spark-based, distributed, in-memory processing platform that provides Spark MLlib, which is used to build ML models in Apache Spark.

It provides major algorithms and tools for classification, clustering, regression, and collaborate filtering. It provides high-level APIs, which are built on DataFrames for constructing, evaluating, and tuning pipelines.

As shown in Figure 7-6, for machine learning workloads, Azure Databricks provides Databricks Runtime for Machine Learning (Databricks Runtime ML), a ready-to-go environment for machine learning and data science. It contains multiple popular libraries, including TensorFlow, PyTorch, Keras, and XGBoost

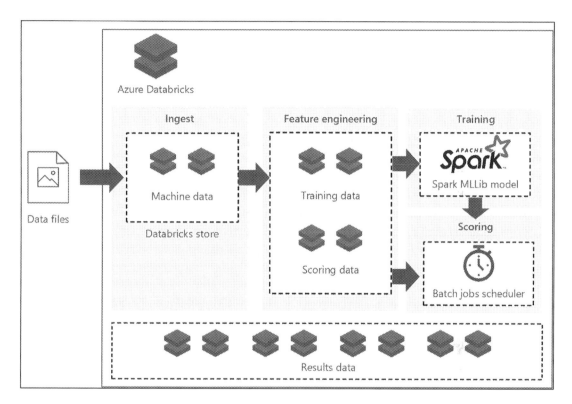

***Figure 7-6.*** *Databricks Spark MLIB ecosystem*

Some of the known features include:

- *Algorithms*: provides algorithms for classification, clustering, regression, and collaborate filtering. The major algorithms include:

  - Basic statistics

  - Regression

  - Classification

  - Recommendation system

  - Clustering

- Dimensionality reduction

- Feature extraction

- Optimization

- *Featurization*: Includes feature extraction, transformation, dimensionality reduction, and selection

- *Pipelines*: Can construct, evaluate, and tune ML pipelines

- *Persistence*: Helps in saving and loading algorithms, models, and pipelines

- *Utilities*: Already built-in utilities for algebra, statistics, and data handling

The advantage of using a Spark MLlib library is that the data scientist can focus on the building the ML models rather than worrying about distributed processing, partitioning of data, infrastructure configurations, optimization, etc. On the other hand, data engineers can focus on distributed system engineering.

Usually, while creating an ML algorithm there is a sequence of tasks that need to be performed, sometimes in a loop. The tasks include preprocessing of data, feature extraction, model fitting, hyperparameter tuning, validation, and the evaluation stage. There are a lot of libraries available for each stage, but making a composite pipeline with a huge set of data is not an easy task.

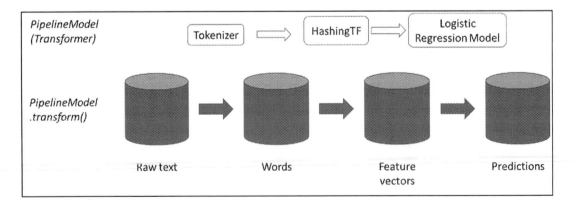

***Figure 7-7.*** *Building a sample ML Model*

Spark makes the job easy here by providing a high-level API for MLlib under the 'spark.ml' package. There are two types of pipelines available: the transformer type (Figure 7-7) and the estimator type. A transformer type takes the dataset as input and provides a processed dataset as output; for example, Tokenizer is a transformer-type pipeline. An estimator type pipeline first fits the input dataset to create a model and then uses this model to transform the input dataset; one example is logistic regression in an estimator type pipeline.

Azure Databricks also supports MLflow, which is an open source platform for managing the end-to-end ML lifecycle. It has the following primary components:

- *Tracking*: Allows you to track experiments to record and compare parameters and results

- *Models*: Allows you to manage and deploy models from a variety of ML libraries to a variety of model-serving and inference platforms

- *Projects*: Allows you to package ML code in a reusable, reproducible form to share with other data scientists or transfer to production

- *Model Registry*: Allows you to centralize a model store for managing models' full lifecycle stage transitions, from staging to production, with capabilities for versioning and annotating

MLflow is supported in Java, Python, R, and through REST APIs as well.

**Azure Notebooks**

Azure Notebooks is a cloud service to develop & run Jupyter Notebooks with zero installation. Jupyter Notebook (Figure 7-8), as we are aware, is an open source project that helps in developing executable code in multiple programming languages like Python, R, and F#. It also provides graphics and visualization in a single canvas.

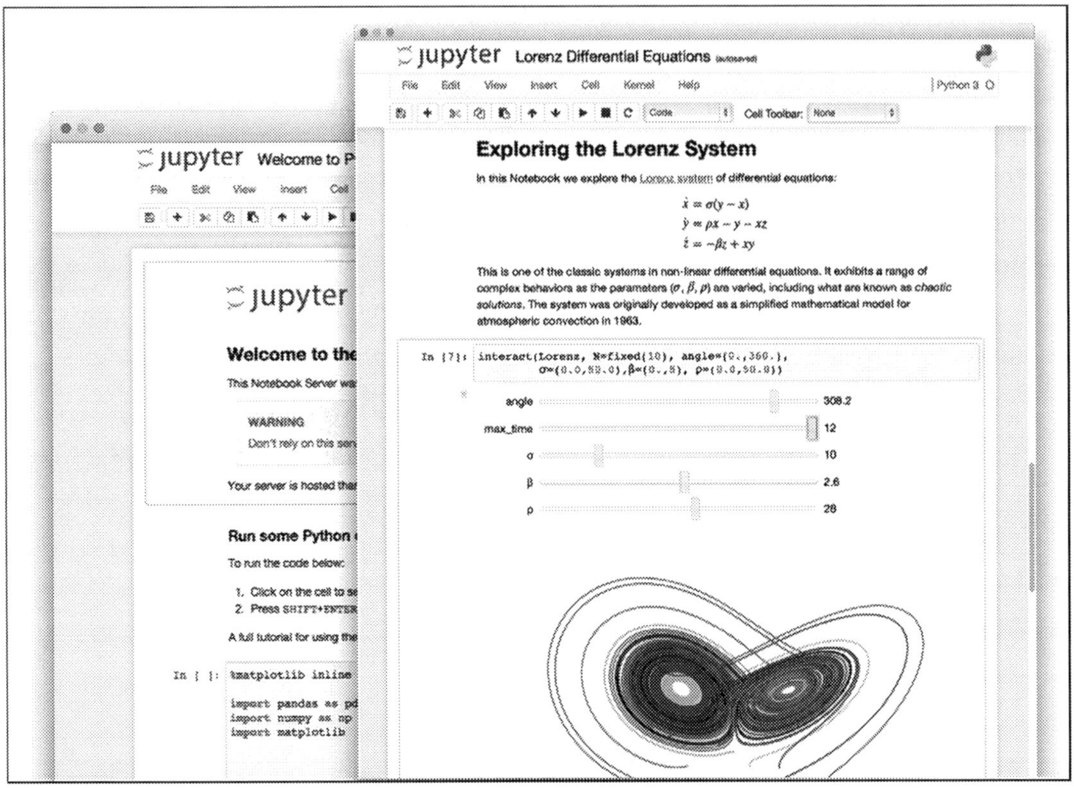

***Figure 7-8.*** *Jupyter Notebook interface*

Jupyter has become popular for many uses, including data science instruction, data cleaning and transformation, numerical simulation, statistical modeling, and the development of ML models.

Azure Notebooks are very useful in prototyping, data science, research, or learning. The advantages of using Azure Notebooks are as follows:

- A data scientist has instant access to a full Anaconda environment with no installation.

- A user has the Python environment available without any hassle.

- A developer can use Notebooks for building a quick code.

Notebooks are very useful as, when required, developers can collaborate on them and share their code, and make rapid changes in an agile manner. The user need not perform any installation or bet concerned about maintaining the environment. Being on the cloud, its' simple to share notebooks with all other authorized users; there are no complications involved with respect to source control of repositories. It's relatively very easy to copy or clone the notebooks for quick experimentations and modifications.

Let's build a sample solution to understand how these components come together to build an advanced analytics solution.

**Exercise: Building a Machine Learning Model in Azure Databricks for Predictive Analytics**

In this exercise we will learn about building an ML model in Azure Databricks to predict whether an upcoming flight will experience any delay or not. It will include performing ETL operations using Azure Databricks. The data is read from a remote file system through Azure Data Factory and stored into Azure Data Lake Storage Gen2. One can further process that data in Azure Databricks for building ML models, summarize data with Azure Synapse Analytics (formally Azure Data Warehouse), and visualize the predictions on a map using PowerBI.

The high-level architecture looks like Figure 7-9.

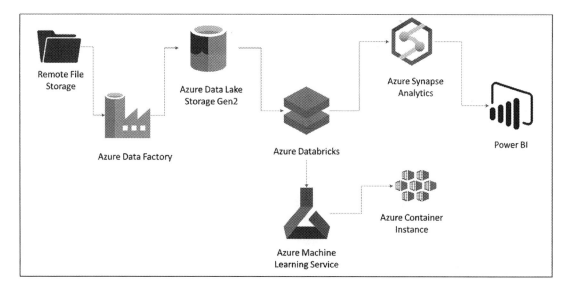

***Figure 7-9.*** *High-level architecture for predictive analytics exercise*

The Azure services required for this exercise include:

1. Azure Data Factory

2. Azure Data Lake Storage (Gen2)

3. Azure Databricks

4. Azure Machine Learning

5. Azure Synapse Analytics

6. Azure Container Instance

7. Power BI Desktop

The exercise includes the following high-level steps:

1. Create an Azure Databricks workspace.

2. Create a Spark cluster in Azure Databricks.

3. Create an Azure Data Lake Storage Gen2 account.

4. Create an Azure Data Factory.

5. Create a Notebook in Azure Databricks to read data from Azure Data Lake Storage, and perform ETL and build ML Models.

6. Create a pipeline in Azure Data Factory to copy data from a remote file system to Azure Data Lake Storage, then invoke Databricks Notebook to perform ETL on the data.

7. Visualization in PowerBI

Let's Start with the first step—create an Azure Databricks workspace:

1. In the Azure portal, create an Azure Databricks workspace by selecting **Create a resource ➤ Analytics ➤ Azure Databricks**.

2. To create the Azure Databricks service, provide the required details, which are

   • Workspace name

   • Subscription

   • Resource group

   • Location

   • Pricing Tier

3.  Select Create, and create the Azure Databricks Service.

Home > Azure Databricks >

## Azure Databricks Service

**\*Basics**    Networking    Tags    Review + Create

**Project Details**

Select the subscription to manage deployed resources and costs. Use resource groups like folders to organize and manage all your resources.

Subscription \*  ⓘ

| Microsoft Azure Internal Consumption | ⌄ |

Resource group \*  ⓘ

| book | ⌄ |

Create new

**Instance Details**

Workspace name \*

| sampledatabricks | ✓ |

Location \*

| West India | ⌄ |

Pricing Tier \*  ⓘ

| Premium (+ Role-based access controls) | ⌄ |

| Review + Create |    | Next : Networking > |

Let's go to the second step—creating a Spark cluster in Databricks:

1.  In the Azure portal, go to the Databricks workspace and select **Launch Workspace**.

2.  You will be redirected to a new portal; there you need to select **New Cluster**.

3.  In the New Cluster page, provide the following details:

    a.  Name of the cluster

    b.  Choose Databricks runtime as 6.0 or higher.

    c.  Select the cluster worker and driver node size from the given list.

    d.  Create the cluster.

Let's go to the third step—creating the Azure Data Lake Storage (Gen2) account:

1.  In the Azure portal, create an Azure Databricks workspace by selecting **Create a resource ➤ Storage ➤ Storage Account**.

2.  To create the Azure Databricks service, provide the required details, which are

- Subscription

- Resource group

- Storage account name

- Location

- Performance

- Account kind

- Replication

- Access tier

Home > New >

# Create storage account

### Project details

Select the subscription to manage deployed resources and costs. Use resource groups like folders to organize and manage all your resources.

| Subscription * | Microsoft Azure Internal Consumption ⌄ |
| --- | --- |
| Resource group * | book ⌄ |
| | Create new |

### Instance details

The default deployment model is Resource Manager, which supports the latest Azure features. You may choose to deploy using the classic deployment model instead.  Choose classic deployment model

| Storage account name * ⓘ | accountsamplestorage ✓ |
| --- | --- |
| Location * | (Asia Pacific) Central India ⌄ |
| Performance ⓘ | ⦿ Standard  ◯ Premium |
| Account kind ⓘ | StorageV2 (general purpose v2) ⌄ |
| Replication ⓘ | Read-access geo-redundant storage (RA-GRS) ⌄ |
| Access tier (default) ⓘ | ◯ Cool  ⦿ Hot |

Review + create          < Previous      Next : Networking >

3.  Next is to enable the Hierarchical namespace under the Advanced tab.

| Basics | Networking | Data protection | **Advanced** | Tags | Review + create |
|--------|-----------|-----------------|--------------|------|-----------------|

**Security**

Secure transfer required ⓘ            ○ Disabled  ⦿ Enabled

Blob public access ⓘ                  ⦿ Disabled  ○ Enabled

Minimum TLS version ⓘ                 [ Version 1.0                                          ⌄ ]

**Azure Files**

Large file shares ⓘ                   ○ Disabled  ○ Enabled

    ❶ The current combination of storage account kind, performance, replication
       and location does not support large file shares.

**Data Lake Storage Gen2**

Hierarchical namespace ⓘ              ○ Disabled  ⦿ Enabled

NFS v3 ⓘ                              ⦿ Disabled  ○ Enabled

    ❶ Sign up is currently required to utilize the NFS v3 feature on a per-subscription
       basis. **Sign up for NFS v3** ⧉

[ **Review + create** ]        [ < Previous ]  [ Next : Tags > ]

4.  Click the Review + create button to create the storage account.

In the fourth step—create an Azure Data Factory Service—do the following:

1.  In the Azure portal, create an Azure Databricks workspace by
    selecting **Create a resource ➤ Analytics ➤ Data Factory**.

2.  To create the Azure Databricks service, provide the required
    details, which are

    •  Name

    •  Version

- • Subscription

- • Resource group

- • Location

- • Enable Git

3. Press the Create button to create.

In the fifth step, you need to create a Notebook in Azure Databricks, which can be used to perform ETL operations, and build and deploy an ML model for predicting any delays in the fight:

1. Open the Azure Databricks workspace and import Python notebooks available at https://github.com/Apress/data-lakes-analytics-on-ms-azure/tree/master/BigData-ML/Notebooks.

2. The notebook is used to perform the following operations:

   a. Prepare the data by performing ETL operations.

   b. Train and evaluate Models.

   c. Deploy the model as web service.

   d. Evaluate the model for batch scoring.

   e. Explore the data to store in Azure Synapse Analytics (formerly SQL DW).

3. The final summary data is stored in SQL DW which is Azure Synapse Analytics.

In the sixth step, you will create a pipeline in the Azure Data Factory service. Do the following:

1. Go to the Azure Data Factory instance created in an earlier step; click Author and monitor.

2. The data files for the exercise are available at https://github.com/Apress/data-lakes-analytics-on-ms-azure/tree/master/BigData-ML/Data.

3. In the Data Factory UI, create a new pipeline.

4. This pipeline is intended to perform the following operations:

    a. Copy data from remote SFTP location to Azure Data Lake Storage Gen2

    b. Invoke the Databricks Notebooks created earlier to perform ETL operations on the data

5. The process looks like the following illustration.

6. To build the preceding pipeline, first you need to use a Copy Data activity that can copy the data from source to sink: source being a remote SFTP server and sink is Azure Data Lake Storage Gen2.

7. Create an SFTP-linked service by passing the following parameters:

    a. Name

    b. Host

    c. Port

    d. Username

    e. Password

8.  Create an Azure Data Lake Storage Gen2-linked service by passing
    the following parameters:

    a.  Name

    b.  Azure subscription

    c.  Storage account name

### New linked service (Azure Data Lake Storage Gen2)

> ⓘ If the identity you use to access the data store only has permission to subdirectory instead of the entire account, specify the path to test connection. Please make sure your self-hosted integration runtime is higher than version 4.0 if connecting via self-hosted integration runtime.

**Name** *

    AzureDataLakeStorage1

**Description**

    
**Connect via integration runtime** *                                      ⓘ

    AutoResolveIntegrationRuntime                                          ⌄

**Authentication method**

    Account key                                                            ⌄

**Account selection method**                                               ⓘ
    ⦿ From Azure subscription      ◯ Enter manually

   **Azure subscription**                                                  ⓘ

      Microsoft Azure Internal Consumption (fc489c93-72d7-4073-8b24-e2f4ea9336f0)   ⌄

   **Storage account name** *

      accountsamplestorage                                                 ⌄

**Test connection**
    ⦿ To linked service      ◯ To file path

**Annotations**
    + New

| Create | Back |          🖉 Test connection | Cancel |

9.  Create an Azure Databricks-linked service by passing the following parameters:

    a. Name

    b. Azure subscription

    c. Databricks workspace

    d. Cluster type

    e. Access token

    f. Cluster ID

**New linked service (Azure Databricks)**

Name *

AzureDatabricks1

Description

Connect via integration runtime *

AutoResolveIntegrationRuntime

Account selection method *

From Azure subscription

Azure subscription *

Microsoft Azure Internal Consumption (fc489c93-72d7-4073-8b24-e2f4ea9336f0)

Databricks workspace *

demoDatabricksService

Select cluster

○ New job cluster    ● Existing interactive cluster    ○ Existing instance pool

Databrick Workspace URL *

https://adb-7321062891290945.5.azuredatabricks.net

( **Access token**    Azure Key Vault )

Access token *

•••••••••••••••••••••••••••••••

Choose from existing clusters *

AdvancedAnalytics

Annotations

**Create**                                    ✐ Test connection    Cancel

10. Configure the Copy data activity by selecting the:

   a. Source dataset through the SFTP-linked service

   b. Sink dataset through the Azure Data Lake Storage Gen2-linked service

| General | Source | **Sink** | Mapping | Settings | User properties |
|---|---|---|---|---|---|

| | | |
|---|---|---|
| Sink dataset * | ▣ DelimitedText1 ⌄ | ✐ Open    + New |
| Copy behavior | None ⌄ | ⓘ |
| Max concurrent connections | | ⓘ |
| Block size (MB) | | ⓘ |
| Quote all text | ✓ | |
| File extension | .txt | ⓘ |

11. Configure the Databricks Notebook activity by:

    a. Selecting the Databricks-linked service created in the earlier step

| General | **Azure Databricks** | Settings[1] | User properties |
|---|---|---|---|

Databricks linked service *    ❋ AzureDatabricks1 ⌄   ✐ Test connection   ✐ Open   + New

    b. Selecting the Databricks Notebook created in the earlier step for performing the ML scoring operation (i.e., deploy for batch scoring)

| General | Azure Databricks | **Settings** | User properties |
|---|---|---|---|

Notebook path *   /Users/▮▮▮▮.com/book/Perf   [📁 Browse]   [ Open ]

▷ Base parameters

▷ Append libraries

12. Once the pipeline is created, save, publish, and trigger the pipeline to monitor the results.

In the seventh step, you will be required to present the data stored in the data warehouse on PowerBI Desktop:

1.  Once the PowerBI is connected to SQL Data Warehouse, by providing the credentials, you will be able to load the data, which would look like the following illustration.

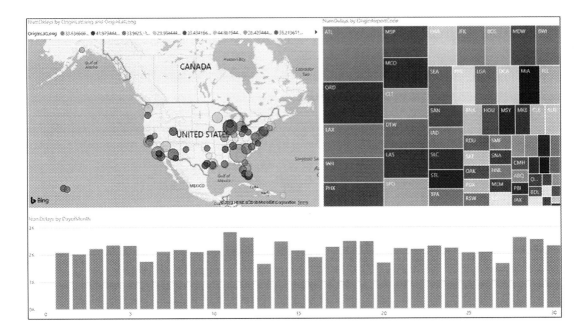

# Summary

In this chapter, basic concepts of data science and various platforms available on Azure were discussed. This book is primarily focused on data engineers, but having a basic understanding of data science really helps to collaborate with other personas in the ecosystem. Microsoft Azure makes the journey to build ML solutions easy, as there are solutions available from beginner to expert-level data scientists—starting from building solutions like cognitive services that need minimal coding to ML designer service where basic level AI/ML solutions can be built. Then Azure Machine Learning service to Apache Spark MLlib and Automated ML can help to build moderate- to advanced-level ML solutions.

# CHAPTER 8

# Model and Serve

Model and Serve is the final phase of a data analytics solution (Figure 8-1). In this phase, transformed data is consumed for the final output, either through visualization or any dependent applications. The entire data journey is planned, based on the target use case. In this chapter, the discussion is on the various scenarios that are applicable in this phase, and how to decide on technologies based on the cost and efficiency.

Data Ingestion and Storage

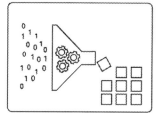

Transfer and store

Data Preparation & Training

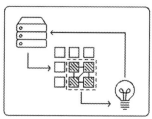

Process and clean

Model & Serve

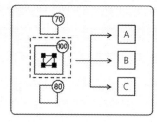

Serve and analyze

***Figure 8-1.*** *Model and serve phase*

## Model and Serve

Several target systems can collect data from the data lake. These systems are typically data warehouses, analytical databases, or data marts. Analytical users may also use operational applications like ERP or CRM and real-time applications. Even data scientists requiring raw data for their models can be the consumers. Let's conceptually discuss these systems:

© Harsh Chawla and Pankaj Khattar 2020
H. Chawla and P. Khattar, *Data Lake Analytics on Microsoft Azure*,
https://doi.org/10.1007/978-1-4842-6252-8_8

- *Data warehouse*: Usually, the data created by the ETL jobs running on the data in a data lake is loaded into the data warehouse by creating files that can be bulk loaded, or by creating simplistic ETL jobs that simply load the data in real time.

- *Operational databases*: An operational database, which could be a relational database or NoSQL DB, is used to organize, clean, and standardize the data. It addresses an ELT approach disadvantage, in which the ELT jobs interfere with the performance of analytics jobs and affect them. Enterprises can protect analytics queries from being slowed down by ELT jobs by moving all processing to a separate operational database.

- *Real-time applications and business intelligence tools*: Incoming data streams are handled by real-time application and BI tools for catering to industry-specific use cases. These applications could also be deployments of ML models created by data scientists in production for statistical analysis. These applications can process data in both batches and in real time.

As this is the final phase in the data analytics solutions, in continuation of the context from previous chapters, there are two major scenarios—based on which, the entire solution is designed:

*Real-time data processing*: Data coming from the prep and train phase will land directly in a data store, where the downstream applications will consume it. Scenarios like stock price prediction where the buy or sell call has to be notified in a short span of time, or IoT scenarios where real-time alerts need to be generated, will fall under these scenarios.

*Batch mode data processing*: Data coming from the prep and train phase will land directly in a data store, and further data modeling will be done. After data modeling, the KPIs will be cached and consumed by data visualization applications like Power BI, Tableau, etc.

Let's explore these scenarios in detail with the help of Figure 8-2.

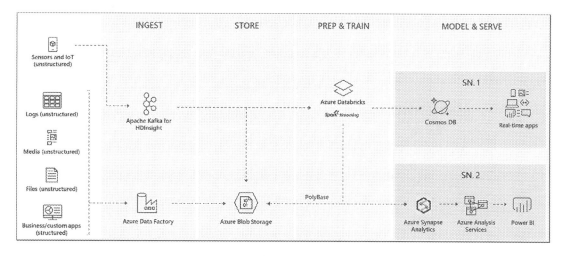

**Figure 8-2.** *Real-time and batch mode data processing*

As per Figure 8-2:

1. Both real-time and batch data is coming from the ingestion layer.

2. Data is getting stored on the Azure Blob storage.

3. Data prep and training is being done with Spark streaming and Databricks for real-time and batch data processing, respectively.

4. Based on the use case:

    a. *Scenario 1 (real-time data refresh)*: The processed data can be either projected directly on dashboarding tools like Power BI or downstream applications to generate alerts or notifications to the end users through Cosmos DB.

    b. *Scenario 2 (batch mode incremental data refresh)*: Data will be further loaded into Azure Synapse Analytics; and build a caching layer is built, using Azure Analysis Services; and Power BI reports are processed from the caching layer.

Let's delve deeper into this phase and understand these scenarios.

# Real-Time Data Refresh

IoT and streaming scenarios—like real-time fleet tracking, real-time fraudulent transactions detection, social media analytics, etc.—generally need real-time data refresh to avert or react to critical events. As discussed earlier, there are two major scenarios under this category:

1. Real time dashboards

2. Real-time downstream applications

**Real time dashboards** – Scenarios like social media analytics, where the real-time impact of a campaign needs to be measured, will need real-time data analysis. Real-time streams coming from event hubs or Apache Kafka can be analyzed using Azure Stream Analytics or Spark streaming. The data from these services can land directly into Power BI or any other supported dashboarding solutions.

With the following exercise, let's understand how the streaming data can be directly projected in Power BI. This is the continuation of same exercise, which was started in Data Ingestion phase, COVID related data was fetched from Twitter and sent to event hub on Azure. In chapter 6, for the prep and train phase, data coming from event hubs was further processed to extract the sentiment of the tweets. Following exercise will help to understand, how the incoming data from Twitter can be directly projected on Power BI to build real time visualizations.

# Stream Analytics and Power BI: A Real-Time Analytics Dashboard for Streaming Data

## Modify the Stream Analytics Job to Add Power BI Output

1. In the Azure portal, open the Stream Analytics job (named twittereventhub) that was created earlier in Chapter 6.

2. On the left menu, select **Outputs** under **Job topology**. Then, select + **Add** and choose **Power BI** from the dropdown menu.

3. Select + **Add** ➤ **Power BI**. Then fill in the form with the details (Figure 8-3) and select **Authorize**.

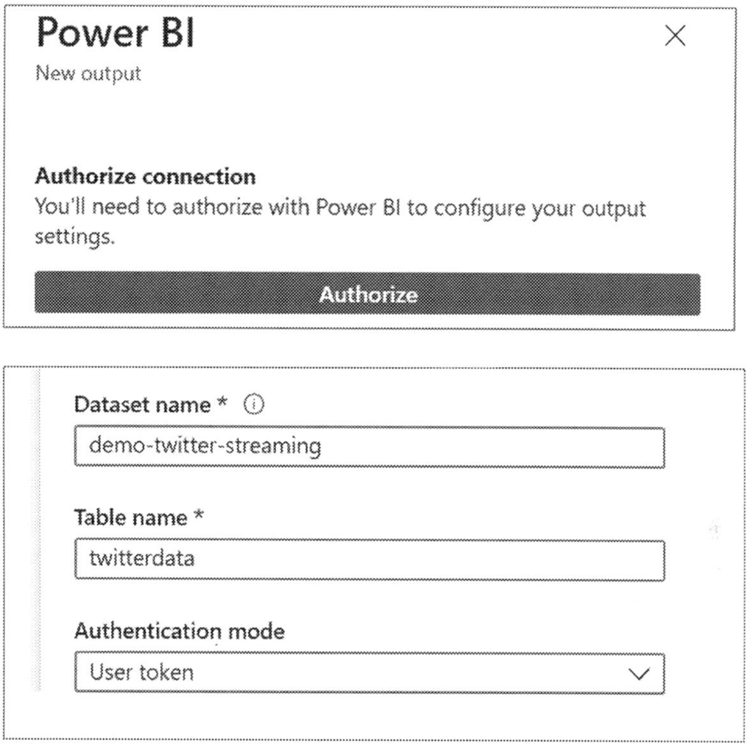

**Figure 8-3.** *PowerBI connection*

4. When you select Authorize, a pop-up window opens up and provides credentials to authenticate to your Power BI account. Once the authorization is successful, Save the settings.

5. Click **Create**.

6. Write and Test the Query.

Once you have defined the dataset for Power BI, you need to define the data that you wish to pull from your Twitter feed.

1. In the Job Topology Section on the left menu, select Query after closing the Output menu (Figure 8-4).

2. You can write your query in the window, which is a simple SQL statement.

For example,

```
SELECT created_at,text
,[user].screen_name
,[user].verified
,[user].location
FROM twitterstream
WHERE 1=1
```

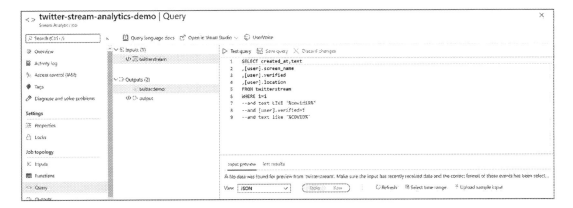

***Figure 8-4.*** *Stream analytics query*

3. Test the Query by pressing the Test Query button and then view the results in the result tab.

4. Run the job.

   a. Make sure your Twitter app is working.

   b. In the Stream Analytics job, hover over the Overview menu and then select Start.

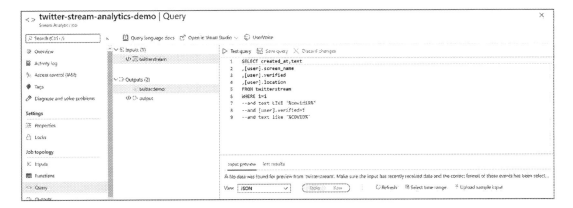

***Figure 8-5.*** *Stream Analytics overview*

Now, the Stream Analytics job fetches all the tweets for COVID-19 and builds the dataset and table for Power BI.

## Create the Dashboard in Power BI

1.  Go to Powerbi.com and sign in with your work or school account. If the Stream Analytics job query outputs results, the dataset will be created (Figure 8-6).

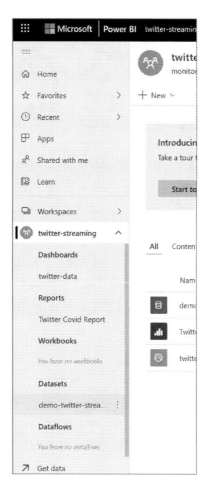

*Figure 8-6.* *PowerBI interface*

*Dataset*: Demo-twitter-stream: define in the Stream Analytics job.

2. In the workspace, click **+ Create**.

3. Create a dashboard.

4. At the top of the window, click **Add tile**, select **CUSTOM STREAMING DATA**, and then click **Next** (Figure 8-7).

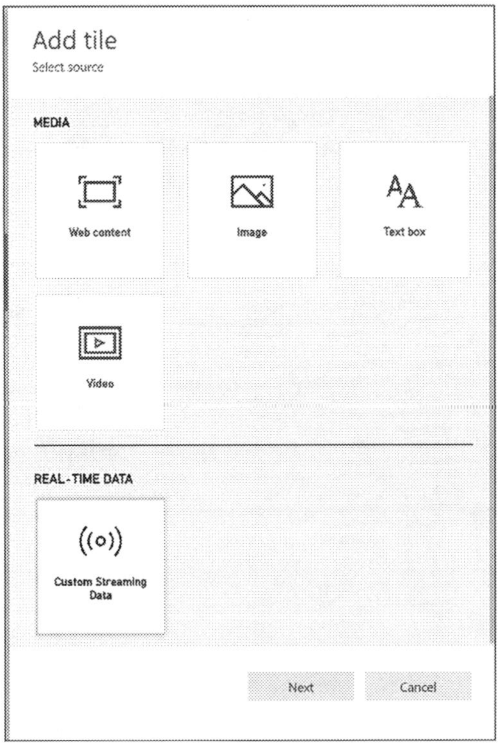

*Figure 8-7. PowerBI custom streaming*

5. Under **YOUR DATSETS**, select your dataset and then click **Next**.

6. Under **Visualization Type**, select **Card**; then, in the **Fields** list, select your field (Figure 8-8).

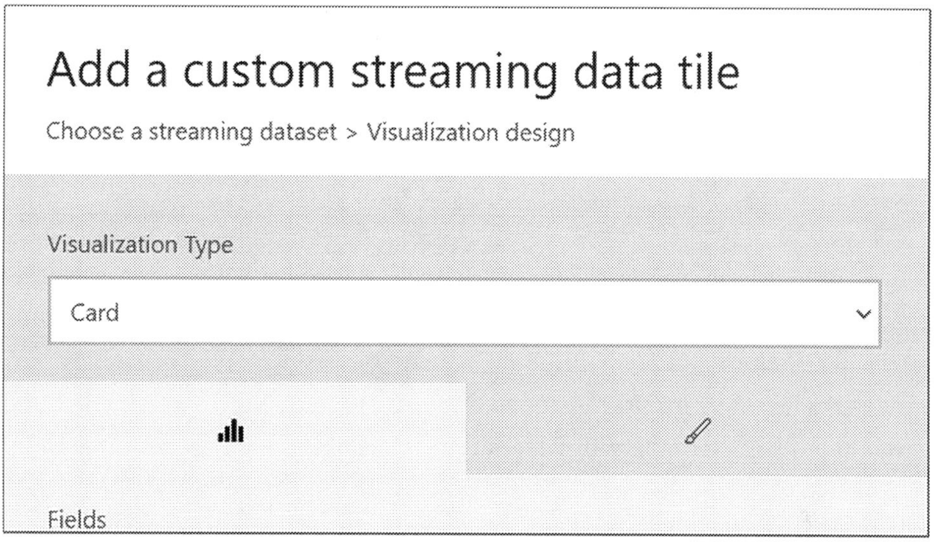

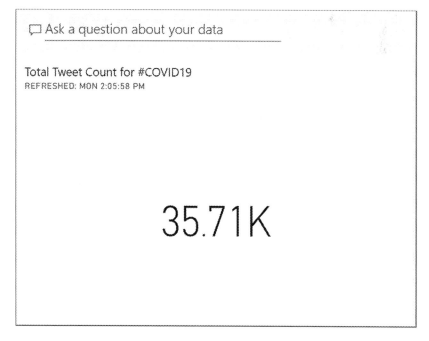

# Add a custom streaming data tile

Choose a streaming dataset > Visualization design

Visualization Type

Card                                                                          ⌄

Fields

*Figure 8-8.* *PowerBI interface*

7.  Click **Next** and fill in the details like a title and subtitle.

8.  Click **Apply**.

9.  Visualization will look like Figure 8-9.

▢ Ask a question about your data

Total Tweet Count for #COVID19
REFRESHED: MON 2:05:58 PM

# 35.71K

*Figure 8-9.* *PowerBI Interface*

**Real-time downstream apps** – Scenarios like stock price prediction and sending buy/sell stock notification to the stock investors, or fleet-tracking scenarios where the application tracks the real-time location of trucks, will need this solution of sending data in real time to downstream applications. IoT data or streaming data comes in the form of JSON files, and the velocity of this input data is very high. Therefore, it's crucial to have a data store that can store data in the JSON format and provide real-time data processing capability with limitless scale. Moreover, where the stakeholders are in different geographic locations, it may need multi-master write capabilities and replicate the changes in real time.

Microsoft Azure offers a solution called Cosmos DB, which caters to the features as mentioned previously. Before getting deeper into these scenarios, let' understand the basics of Cosmos DB.

**Cosmos DB** – Azure Cosmos DB is a globally distributed, multimodel NoSQL database service. It can infinitely scale across any number of Azure regions worldwide. This is a serverless data store with low order-of-millisecond response time. It supports multiple data models (key-value, documents, graphs, and columnar) and multiple APIs for data access, including Azure Cosmos DB's API for MongoDB, SQL API, Gremlin API, and data models API natively, and in an extensible manner.

Here are a few features that make it stand out:

- *Multiple data model and API support*: Key-value, column-family, document, and graph database models and APIs are supported.

- *Turnkey global distribution and multi-master writes*: Cosmos DB has global distribution capability, which can be enabled with a click of a button and can replicate data in any Azure datacenter. Moreover, it supports writes from multiple replicas and seamlessly replicates these changes across all the DB replicas.

- *High availability and tunable throughput*: Azure Cosmos DB offers 99.999% guarantees for high availability for both reads and writes. It can elastically scale throughput and storage and provides low latency for reads and writes.

- *Change feed*: Cosmos DB is used for event sourcing to power event-driven architectures, using its change feed functionality. Change feed support in Azure Cosmos DB works by listening to an Azure

Cosmos container for any changes. The changes are persisted, can be processed asynchronously and incrementally, and the output can be distributed across one or more downstream microservices and consumers for parallel processing.

Let's discuss some of the concepts that are important to understand before spinning up Cosmos DB:

1. Request Units

2. Database

3. Container

4. Partition key

5. Query API

6. Consistency level

**Request Units** – The basic unit of throughput in Cosmos DB is called a Request Unit (RU). RUs are calculated depending upon the operations done on a document, its size, consistency level, indexing policy, etc. There is a tool called Cosmos DB capacity calculator that helps to identify the correct RUs, based on the type of operations to be performed in an application.

**Database** – A database is a collection of containers, which further consists of actual data. RUs can be provisioned at the database level, which can further be shared by all the containers in a database.

**Container** – A container is a collection of actual data. RUs can be provisioned at a container level for dedicated use or can be shared from the RUs from DB.

**Partition key** – A partition key helps to ensure data is segregated across multiple partitions, and helps to scale horizontally. While creating a database and container, this is a mandatory field to add a partition key. The partition key should be based on a column or composite columns with the highest number of unique values. This will ensure even distribution of data across multiple partitions.

**Query API** – There are five supported APIs and respective data models in Cosmos DB:

1. **SQL** – This is the most preferred API in Cosmos DB to store JSON documents. This supports server-side programming like stored procedures, triggers, functions, etc.

2. **MongoDB** – This is another document store data model available in Cosmos DB. MongoDB APIs can be used to migrate data from MongoDB into Cosmos DB. It helps to migrate the applications into Cosmos DB without any major code changes.

3. **Cassandra** –Cassandra API can used to store columnar data. This API also supports seamless data migration into Cosmos DB and provides Cassandra query language support.

4. **Azure Table** – This is a key-value pair storage, and Cosmos DB supports this store and API to fetch the data.

5. **Gremlin (graph)** – This API can be used to store graph data, and it provides seamless migration of the application using Gremlin with minimal code change.

**Consistency levels** – Cosmos DB supports five consistency levels: eventual, consistent prefix, session, bounded-staleness, and strong. As shown in Figure 8-10, by choosing an appropriate consistency level, programmers can choose between consistency and availability of data. As discussed in Chapter 1, NoSQL DBs are based on the CAP theorem. This is the practical implementation of the same concept.

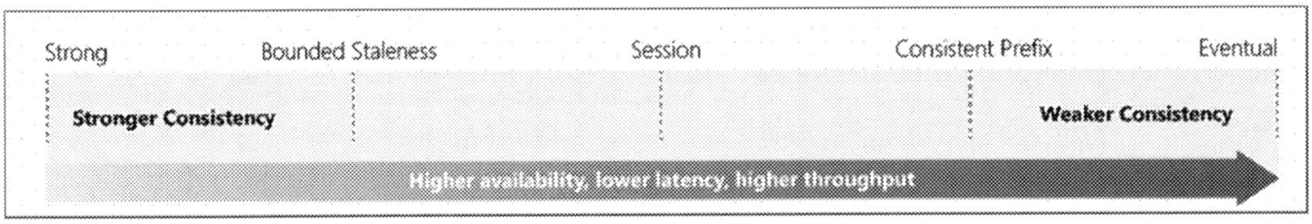

| Strong | Bounded Staleness | Session | Consistent Prefix | Eventual |
| --- | --- | --- | --- | --- |
| **Stronger Consistency** | | | **Weaker Consistency** | |

Higher availability, lower latency, higher throughput

*Figure 8-10.  Consistency levels in Cosmos DB*

By now, the basics of Cosmos DB have been covered. Further to that, let's explore how Cosmos DB's change feed and recently added HTAP capabilities can be helpful in the model and serve phase. These two features are discussed as follows:

1. *Change feed*: Change feed in Cosmos DB tracks all the changes done in a container. It then sorts the list of documents in the order in which the change was applied. This feature gives lots of flexibility to notify the users using Azure functions, invoke event hubs, or store data directly into the storage layer (Figure 8-11).

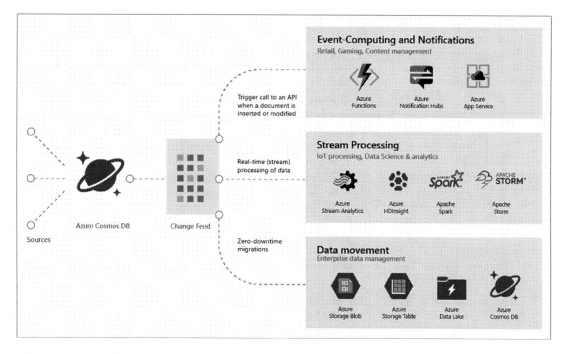

***Figure 8-11.***  *Change feed scenarios*

2.   This feature can be useful in IoT scenarios to raise alerts
for a specific event like temperature rising above a certain
threshold, or notifying the end users if part of some machinery
is malfunctioning. Moreover, there are applications that directly
write to Cosmos DB and may not need the complete lifecycle of a
data analytics solution. Scenarios like retail, where a customer's
shopping data is stored, can use this change feed functionality
to understand what items were removed or added to the cart.
Accordingly, product recommendations can be made to upsell
more products or maybe pass some discounts to get a successful
purchase.

Change feed can be read by either a push or pull mechanism. Under push,
there are two ways to invoke further actions: Azure functions or change feed
processor. For pull, only the change feed processor can be used. In the Cosmos
DB console, there is an option to select Azure functions. When the Azure
functions option is clicked, the following (Figure 8-12) is invoked; check various
output options to process the change feed.

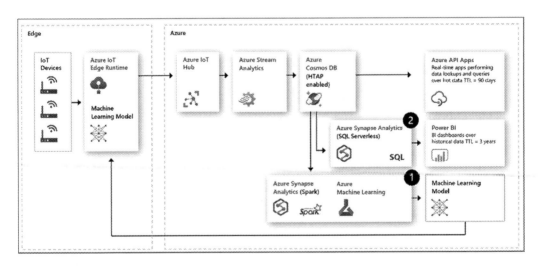

**Figure 8-12.** *Azure functions output options*

However, Azure function service should be created prior to clicking this option.

3. *HTAP scenario*: As shown in Figure 8-13, real-time IoT data is coming through the stream analytics job and landing in Cosmos DB. Using the analytical store feature, this data can be replicated to Synapse sink and can be processed further using Spark notebooks or SQL queries.

**Figure 8-13.** *IoT predictive maintenance scenario*

This is going to be a great and well-used feature; it's in preview right now. This feature ensures the data is replicated into a Synapse Analytics sink for further analytics without the need of ETL/ELT processing. This feature will open doors to build predictions using Spark MLlib, or merge this data with other data and build dashboards in Power BI. To enable this feature, the following steps need to be followed (Figure 8-14):

1. Enable Synapse link in Cosmos DB under features setting

2. Set Analytics store to "ON" while creating a new container

3. Click ok to create the container.

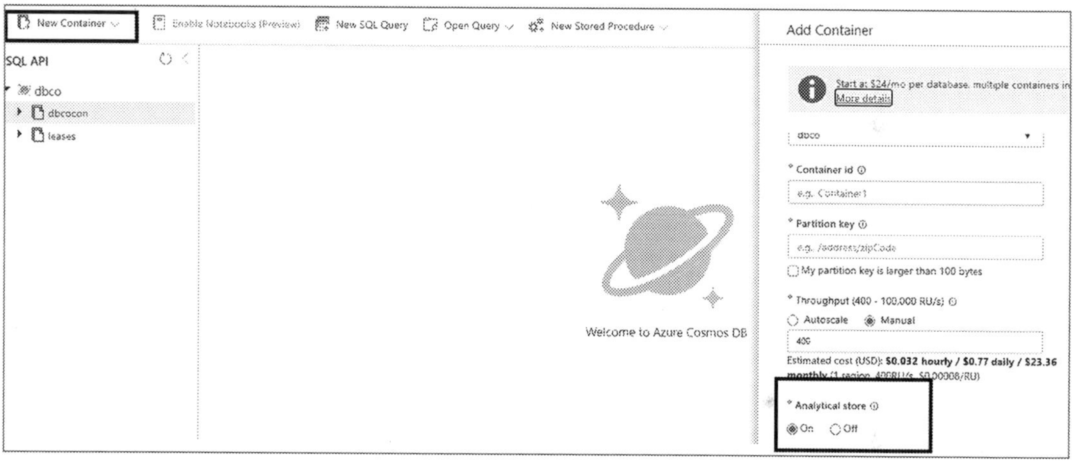

***Figure 8-14.*** *Enable analytical store*

4. Go to the Synapse workspace and Click the Data tab.

5. Under the Linked tab, you can add the Cosmos DB database by adding the following information:

   a. Select **Connect to external data**

   b. Select the API (SQL or MongoDB).

   c. Select ***Continue.***

   d. Name the linked service.

   e. Select the **Cosmos DB account name** and **database name**.

f.  (Optional) If no region is specified, Synapse runtime operations will be
routed toward the nearest region where the analytical store is enabled.

g.  Select **Create**

After the Cosmos DB is linked to Synapse, the data is automatically replicated under
Cosmos DB storage. After the data is replicated, Spark notebooks or SQL queries can be
used to write code for the further processing of data, as shown in Figure 8-15.

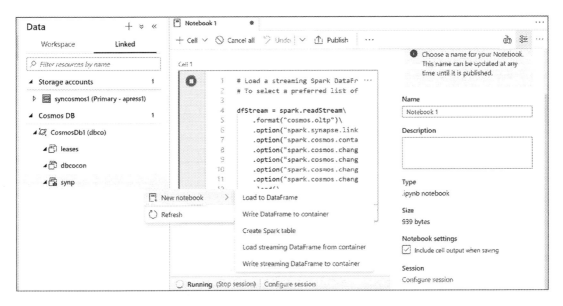

***Figure 8-15.***  *Query Cosmos DB data in Synapse workspace*

This capability of Cosmos DB makes data analytics extensible and builds multiple
layers using Synapse Analytics capabilities.

In summary, under the model and serve phase of a data analytics solution, new
features of Cosmos DB, Synapse Analytics, Spark ML, and Power BI provide lots
of features and flexibility for IoT, retail, ecommerce, etc. Data can be either shown
directly in a dashboard or used to trigger another data pipeline and leverage ML and
Spark, which can be used to enhance real-time response of an application. With these
technologies the possibilities are endless, and data can be transformed into various
forms to extract maximum output.

Now, let's get into the model and serve scenarios of data coming from batch mode
data processing.

**Batch mode data processing** – Under this scenario, the data comes from disparate data sources. As shown in Figure 8-16, data can come from data sources like business applications, flat files, media, or even IoT sensors, and it can be structured, semistructured, or unstructured. This data can be put into data lake storage using an ETL service like Azure Data Factory or Informatica. After the data is moved to the data stores, Apache Spark can be used to prepare and train the data. Finally, this data is ready to be modeled and served.

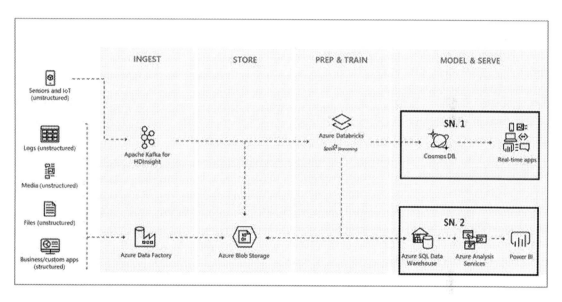

***Figure 8-16.*** *Data analytics solution architecture before Synapse Analytics*

Scenario 1, for the data coming in real-time, has been discussed in the preceding section. Now, let's discuss scenario 2: batch mode data processing. In this scenario, generally the delay of a few hours is acceptable. ELT jobs bring the data at regular intervals, and the transformation engine kicks in accordingly during the day and puts the data in the model and serve phase.

In the model and serve phase, under batch mode, the most preferred technology is Azure Synapse Analytics (formerly called SQL DW). By the time the data is transformed, there is always a structure to it, and putting it in a structured data store is always a preferred choice. However, it can go into any other preferred technology solution under this category.

If Azure Synapse Analytics (SQL pool) isn't needed to run 24/7, Azure Analysis Services comes into the picture; it acts as a caching layer, which needs to be refreshed multiple times during the day. The entire BI and dashboarding can be served from this layer, and Azure Synapse Analytics can be activated on-demand to refresh the cache. However, there can be scenarios where the SQL pool needs to run 24/7.

Azure Synapse Analytics is much more than an SQL DW (MPP) offering; it has got Spark, SQL MPP, and orchestration under one workspace called Azure Synapse Analytics workspace. Let's understand a little more about this technology solution in the following section.

## Azure Synapse Analytics

Azure Synapse is a limitless analytics service that integrates data warehousing, big data, and data integration in one place. This capability has disrupted the way data analytics and advanced data analytics solutions could be built. Let's understand how the solutions were built earlier and how these are anticipated to change with the new features in Azure Synapse Analytics.

Referring to Figure 8-16, typical steps in a data analytics solution are:

1. Ingest data from different sources using Azure Data Factory (ADF).

2. Store the data into Azure Data Lake Storage or Blob storage.

3. Use Spark to crunch and transform the data.

4. Apply ML using Spark ML or other ML services.

5. Load the data in SQL DW using ADF.

6. Build cubes and dimensions in Azure Analysis Services.

7. Build dashboards using Power BI on top of Azure Analysis Services.

With Synapse Analytics workspace's capabilities, the solutions may change as follows:

1. Ingest data from different sources using pipeline feature under Synapse Analytics workspace.

2. Store the data into Azure Data Lake Storage or Blob storage.

3.  Create Spark pools to crunch and transform the data using Spark notebooks.

4.  Apply Spark ML using Spark notebooks in the same workspace.

5.  Pick up the data in SQL pools using PolyBase with an SQL query (no need to manually insert data again using ADF).

6.  Build cubes and dimensions in Azure Analysis Services outside the Synapse workspace. This step is optional but can save lots of money, as there wouldn't be any need to keep Synapse Analytics SQL pool running all the time.

7.  Build dashboards using Power BI on top of Azure Analysis Services.

Technically there is no change, but now everything could be done under a single Azure Synapse workspace without manually moving data between different components.

Currently, Synapse Analytics is in public preview, and multiple changes are expected until it's generally available. However, let's explore a little more about these components. There are the following four components under Synapse Analytics workspace:

1.  *Synapse SQL*: Complete T-SQL–based analytics—Generally available (GA)

    a.  SQL pool (GA)

    b.  SQL on-demand (preview)

2.  *Spark*: Deeply integrated Apache Spark (preview)

3.  *Synapse pipelines*: Hybrid data integration (preview)

4.  *Studio*: Unified user experience (preview)

**Synapse SQL** – Synapse SQL is an MPP offering under the Synapse workspace, and it's simply a new name for Azure SQL Data Warehouse. There are two options available to spin up this service:

1.  *SQL pool*: It's a provisioned Azure SQL DW instance that will be continuously running to process the queries. It's charged based on the data warehouse units under the Compute optimized Gen1 tier and units under the Compute optimized Gen2 tier.

2.  *SQL on-demand*: This is a new offering which can be spun up on demand. It's for ad hoc analytics and data transformation requests. It's charged based on the TBs of data scanned in the queries and is a serverless service.

Moreover, there is an option to select either an SQL on-demand or SQL pool instance before executing the query.

**Apache Spark** – This is another capability available in the Azure Synapse Analytics workspace. This is a service that can be provisioned as an Apache Spark pool. This service can be spun with a minimum of three nodes (i.e., one driver and two workers). Spark notebooks can be written on this Spark cluster, and this data can be picked up SQL using PolyBase seamlessly.

**Synapse pipelines** – This is an orchestration engine that is like ADF and is part of the Synapse Analytics code base. In fact, the interface is also like ADF, and the concept of linked services and pipelines, etc. is also the same.

**Synapse Studio** – This is a common platform to perform all the aforementioned activities. This is an interface that gives flexibility to navigate to multiple options, as shown in Figure 8-17.

***Figure 8-17.*** *Synapse workspace interface*

**Data** - Under data, all the database, storage accounts, and even Cosmos analytics data falls under this option.

**Develop** – Under develop, the option to create SQL script, Spark notebook, Data flow, or Spark job definition can be created.

**Orchestrate** – Under orchestrate, the ETL/ELT pipelines can be configured. It's a similar canvas as ADF has.

**Monitor** – Under monitor, pipelines, SQL script, and Spark jobs can be monitored.

**Manage** – Under manage, all the resources created under this workspace can be managed; even security can be configured under this option.

The coming sections discuss the other technologies like Azure Analysis Services and Power BI, under the model and serve phase/

## Azure Analysis Services

Azure Analysis Services (AAS) is a fully managed platform as a service (PaaS) that provides enterprise-grade data models in the cloud. It supports Analysis Services tabular models, which enables any tabular models to be moved into Azure.

Using data from multiple data sources, defining and calculating metrics, and securing your data in a single, trusted tabular semantic data model is enabled by using advanced mashup and modeling features that AAS has to offer. The data model provides an easier and faster way for users to perform ad hoc data analysis, using tools like Power BI and Excel. In the traditional world, SSAS was used for cached reporting. AAS offers this very experience and more for the cloud.

Azure Analysis Services lets you govern, deploy, test, and deliver your BI solution. It is the engine powering the Power BI Premium platform.

Features that stand out for modeling:

- Enterprise grade analytics engine with built-in schema

- *Create and Deploy*: Tabular models (1200 compatibility) including Direct Query, Partitions, Perspectives, RLS, and Translations

- *Manage*: Fully managed PaaS; up to 100 GB memory; SSMS & XMLA management APIs; pause and resume; elastic scale up/down (GA); 99.9% uptime SLA (GA)

- *Consume*: Interactive query performance over large datasets; simplified view over complex data; data models; single model for one version of the truth; full MDX and DAX support. Offers perspectives to define viewable subsets of model focused for specific business reporting

- *Easy integration with Azure services*: Secure role-based data access through integration with Azure Active Directory (AAD). Integrate with ADF pipelines for seamless controlled model development and data orchestration. Lightweight orchestration of models using custom code can be leveraged with the Azure Automation and Azure functions.

- *DevOps friendly*: Version control; scale out for distributed client query

## Power BI

Power BI is the business intelligence offering from Microsoft that delivers beyond the quintessential dashboards and visualizations. The Power BI platform is a collection of software services, apps, and connectors that work together to give coherent, visually immersive and interactive insights out of your varied data sources.

The Power BI platform comprises three apps: Power BI Desktop, Power BI Pro, and Power BI Premium. A citizen user can now do data analysis, modeling, and reporting, which were originally supposed to be done only by business analysts or data scientists.

A Power BI dashboard is a single page, often called a canvas, which tells a story through visualizations.

**Power BI Desktop –** It's a free version that can be installed on your desktop. In spite of being free, there is a lot to offer:

- Connectors to 70 cloud-based and on-premises sources

- Rich widgets and visualizations

- Storage limit of 10 GB per user

- Data relationships between tables and formats autodetected

- Export your reports to CSV, Microsoft Excel, Microsoft PowerPoint, and PDF

- Save, upload, and publish to the Web and Power BI service

Limitations:

- No app workspaces; no API embedding

- No email subscriptions

- No peer-to-peer sharing

**Power BI Pro** – This is the upgraded version of Power BI Desktop, which allows you to create unlimited dashboards, reports, and share across the organization. Power BI Pro has one-user licensing.

Features that are added in Power BI Pro are:

- Ability to embed Power BI visuals into apps (PowerApps, SharePoint, Teams, etc.)

- Share datasets, dashboards, and reports with other Power BI Pro users

- Can create app workspaces

- Peer-to-peer sharing

- Native integration with other Microsoft solutions (Azure Data Services)

**Power BI Premium** – For large organizations that require sharing and hosting of datasets across the organization without individual licensing, Microsoft offers Power BI Premium, a hosted service (SaaS). Power BI Premium provides dedicated hardware and capacity hosted by Microsoft for reliable and consistent performance. Power BI Premium includes several other workloads, including dataflows, paginated reports, and AI—apart from datasets. You have an ability to scale up and down the service tiers as per consumptions.

Features with Synapse and AAS help serve the data that is modeled and visualized:

- *Import*: Import the data into Power BI for small data sources and personal data discovery.

- *Direct query*: Connect directly to Synapse in order to avoid data movement and delegate the query movement to the back-end source there by leveraging its compute power.

- *Composite models and aggregation tables*: Keep summarized data local within the Power Bi Premium workspace and get detail data from the source. Composite models are individual datasets that support tables with direct query, import, or both—all in a single dataset.

- Prep and model your data with ease. Save time and make data prep easier with data modeling tools. You can ingest, transform, integrate, and enrich data in Power BI. Supports self-service Power Query

- Ability to embed Power BI visuals into apps (PowerApps, SharePoint, Teams, etc.)

- Larger storage sizes for extended deployments compared with Pro licensing

Apart from the aforementioned options, there is another unique offering recently launched on Microsoft Azure called Azure Data Explorer. Let's understand how Azure Data Explorer can disrupt the way data analytics solutions are built.

# Azure Data Explorer

Interactive analytics on terabytes and petabytes of data is a challenging job. Azure Data Explorer (ADX; Figure 8-18) helps to solve this challenge well by analyzing billions of rows in seconds. ADX is an analytical service that customers can use for doing real-time analytics on data that is streaming from IoT devices, applications, websites, and other sources. It includes a rich query engine that is optimized for low latency in JSON and lightning flush query performance. It's a fully managed Cloud Service.

Users can take advantage of this service for building SaaS applications, IoT applications, and to do time series analytics on them or build solutions around analyzing log and telemetry data at a really large scale. The system is optimized for querying over structured data, semistructured data like JSON and XML, and unstructured data like free text. Data ingesting is also very easy, as users can ingest data through both streaming and batch sources.

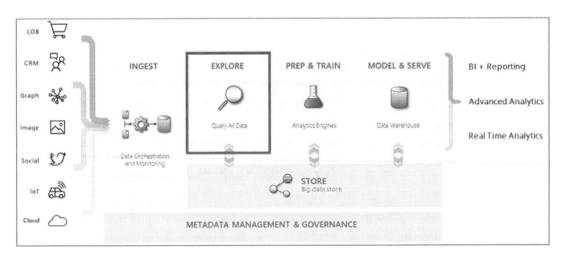

***Figure 8-18.***  *Explorer phase*

Hence it gives users the possibility to handle different scenarios in the same database. So, you don't have to query one database for searching data and another database for aggregations; one can perform everything in one database (i.e., search and also run analytical queries).

ADX provides native connectors to Azure Data Lake Storage, Azure SQL Data Warehouse, and Power BI and comes with an intuitive query language so that customers can get insights in minutes.

Intended for speed and effortlessness, ADX is architected with two services: The Engine and Data Management (Figure 8-19).

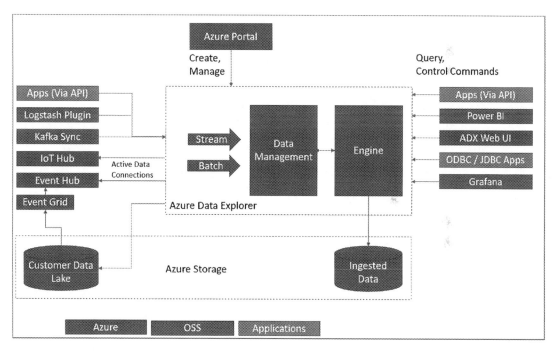

***Figure 8-19.***  *Azure Data Explorer architecture*

The Data Management (DM) service ingests different sorts of crude information and oversees failures, backpressure, and performs information prepping tasks. The DM service likewise empowers quick information ingestion through techniques like autoindexing and compression.

The Engine service is liable for handling the incoming raw information and serving client queries. It utilizes a mix of autoscaling and data sharding to accomplish speed and scale. The read-only query language is intended to make the syntax simple to peruse, write, and automate. The language gives a characteristic movement from one-line questions to complex information handling contents for proficient query execution.

Work in Azure Data Explorer generally follows this pattern:

1. *Create database*: Create a cluster and then create one or more databases in that cluster.

2. *Ingest data*: Load data into database tables so that you can run queries against it. Data ingestion could be in batch or streaming mode.

3. *Query database*: Use our web application to run, review, and share queries and results. It is available in the Azure portal and as a stand-alone application. In addition, you can send queries programmatically (using an SDK) or to a REST API endpoint. ADX provides Kusto Query Language (KQL) for querying the data.

Some of the most prominent functions in Kusto Query Language (KQL) are listed in the following table.

| Operator/Function | Description |
| --- | --- |
| **Filter/Search/Condition** | **Find relevant data by filtering or searching** |
| where | Filters on a specific predicate |
| search | Searches all columns in the table for the value |
| take | Returns the specified number of records. Use to test a query |
| case | Adds a condition statement, similar to if/then/elseif in other systems |
| distinct | Produces a table with the distinct combination of the provided columns of the input table |
| **Date/Time** | **Operations that use date and time functions** |
| format_datetime | Returns data in various date formats |
| **Sort and Aggregate Dataset** | **Restructure the data by sorting or grouping them in meaningful ways** |

(*continued*)

| Operator/Function | Description |
|---|---|
| sort | Sorts the rows of the input table by one or more columns in ascending or descending order |
| count | Counts records in the input table (for example, T) |
| join | Merges the rows of two tables to form a new table by matching values of the specified column(s) from each table. Supports a full range of join types: flouter, inner, innerunique, leftanti, leftantisemi, leftouter, leftsemi, rightanti, rightantisemi, rightouter, rightsemi |
| union | Takes two or more tables and returns all their rows |
| range | Generates a table with an arithmetic series of values |
| **Visualization** | **Operations that display the data in a graphical format** |
| render | Renders results as a graphical output |

Figure 8-20 shows the ADX application with cluster added and a query with result.

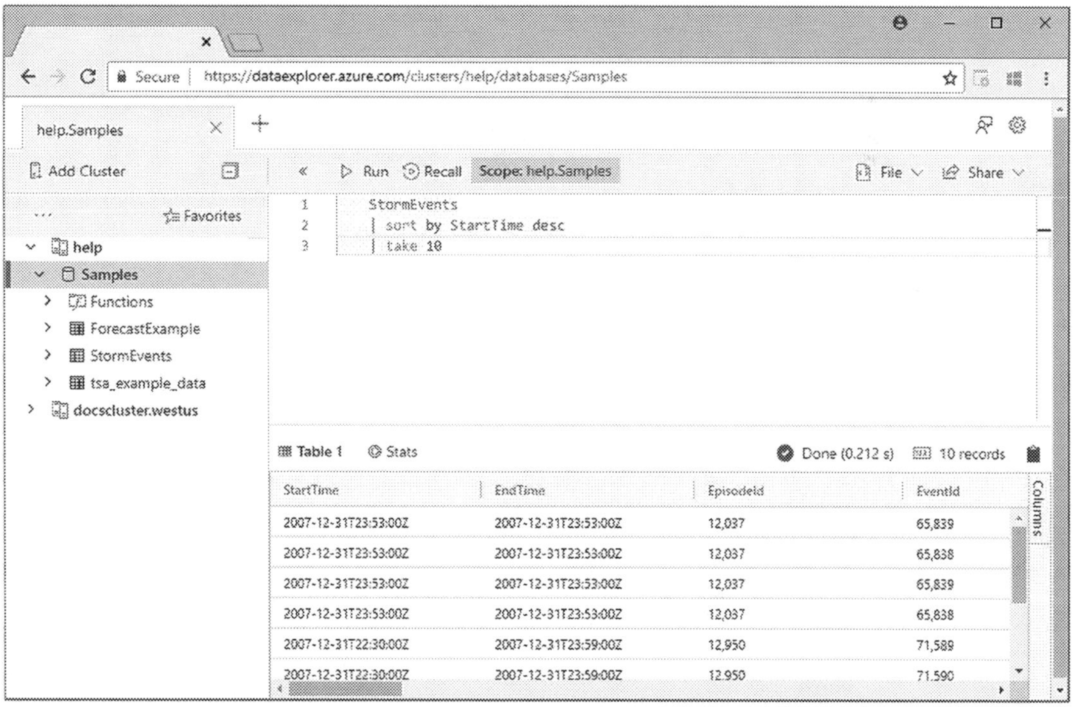

***Figure 8-20.*** *Azure Data Explorer interface*

The queries can also be shared (Figure 8-21).

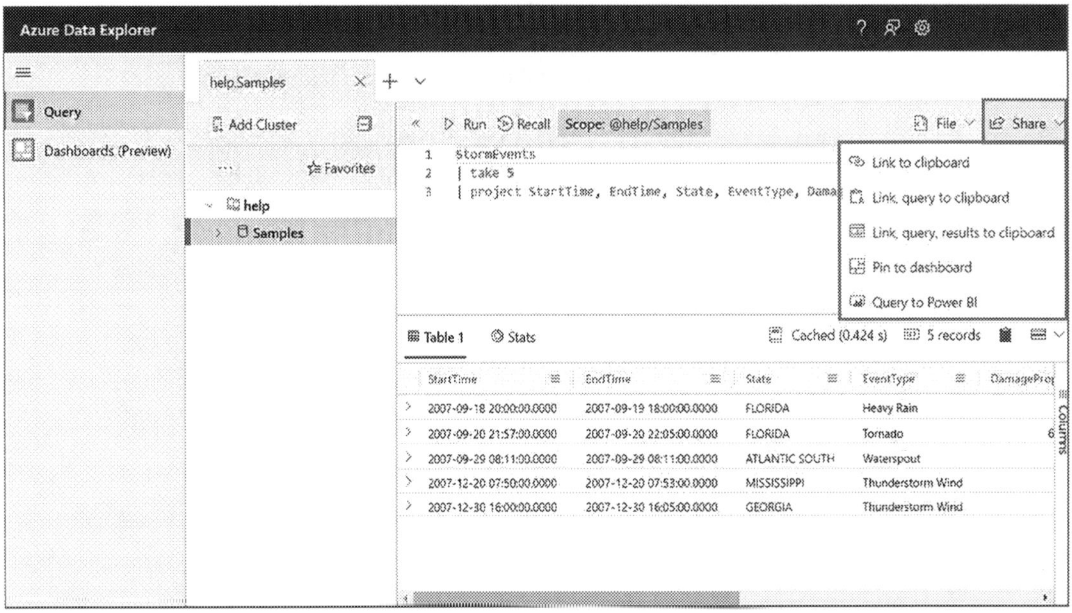

**Figure 8-21.** *Azure Data Explorer interface*

The following options are available in the drop-down:

- Link to clipboard

- Link query to clipboard

- Link, query, results to clipboard

- Pin to dashboard

- Query to Power BI

So, in summary, why should you use Azure Data Explorer (ADX)? Well, because ADX is a fully managed data analytics service that can be embedded in SaaS applications to build multi-tenant or single-tenant SaaS analytics solutions for time series logs, events, transactions, and security data. No infrastructure management or maintenance is required. You create and manage the cluster using Azure Portal. ADX supports advanced analytics, geospatial analytics, and model training or you can use your own model for: scoring of time series data, anomaly detection, forecasting, regression, and to incorporate inline Python and R. ADX has a query editor with powerful intelligence and built-in visualization, support, and sharing and ADX provides deep integration with popular dashboard tools. There is a rich connecting ecosystem that allows you to easily create business workflows, bi-directional data, and lake integration with external tables that enable you to provide a unified query experience and achieve deeper insight and continuous export of information assets. ADX is a big data platform that supports near-real-time insight over petabytes of data. ADX supports high-throughput ingestion and includes a streaming mode with sub-second response time for queries spanning billions of records. ADX instantly analyzes freshly ingested data in its raw format, and ADX can automatically scale as your business grows or shrinks.

## Summary

This chapter explored the options like Cosmos DB HTAP capabilities, Synapse Analytics workspace, and Azure Data Explorer that are completely new capabilities added to Microsoft Azure. These technologies are going to accelerate the adoption of data analytics solutions multifold. The model and serve phase deals with data modeling and consumption, and the previously discussed options make it easy to build and consume. Not only useful for dashboards, these features can also be used to further trigger events, feed data to other applications in real time, and make these solutions of endless scale.

# CHAPTER 9

# Summary

Congratulations for reading this entire book and reaching this summary section. In case, the reader is a data engineer and is trying to learn the technology stack on Microsoft Azure, this book could be an excellent source of information on various architecture patterns and services that could be consumed to build data pipelines. However, if the reader is a beginner into the field of data engineering, this book has lots of exercises along with the theory about various options, which the reader can explore to get hands-on experience on Azure services.

The content of this book has been written considering two personas: data professionals working on relational stores, and data professionals working on nonrelational stores. Since the journey of data varies in both cases, and if the next target is to become a data engineer, learning cross-technologies could be mind boggling. Content in this book has been designed keeping both the personas in mind. In fact, both the writers of this book have been associated with relational and nonrelational fields. This has helped to write content that can cater to both personas.

Let's summarize this book in a nutshell. First, understanding relational and nonrelational data coming in real time and batch mode is important. The journey of this data is completely different in data analytics solutions. In the first three chapters, the major focus was to set the context. The key takeaways from the first three chapters are as follows:

1. There are two major architecture patterns in data analytics solutions (i.e. modern data warehouses and advanced data analytics).

**Modern data warehouse** – It's consolidation of both enterprise data warehouse and big data analytics. The architecture of a modern data warehouse is shown in Figure 9-1.

H. Chawla and P. Khattar, *Data Lake Analytics on Microsoft Azure*,
https://doi.org/10.1007/978-1-4842-6252-8_9

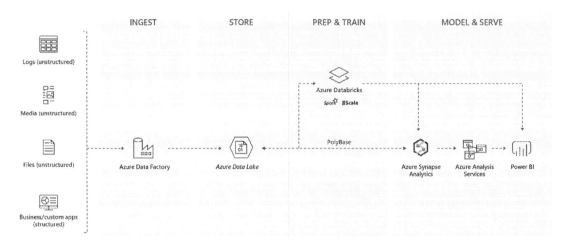

***Figure 9-1.*** *Modern data warehouse*

Modern data warehouses can analyze structured, unstructured, and semistructured data in a single platform.

**Advanced data analytics** – Advanced data analytics is infusion of ML into modern data warehouses along with real-time stream processing. The architecture of advanced data analytics looks like Figure 9-2.

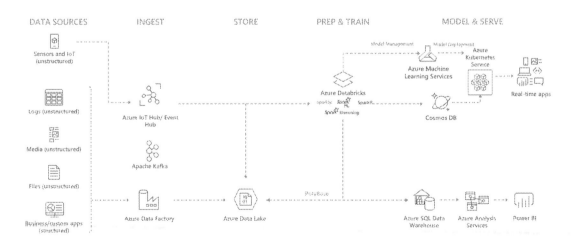

***Figure 9-2.*** *Advanced data analytics architecture*

Advanced data analytics deals with building predictions and prescriptions using ML technologies. The ML models can be consumed in real time or on batch mode data for scenarios like detecting real-time frauds, predictive maintenance, customer churns, or even raising proactive alerts.

Moreover, this book is about building these solutions on Microsoft Azure. There has been in-depth discussion on most of the relevant services, to build this platform.

In the end, there are four key phases of a data analytics solution and respective services on Azure, as follows:

1. Data Ingestion ➤ Event hubs, IOT hubs, Apache Kafka

2. Data Store ➤ Azure Data Lake Storage and Azure Blob storage

3. Prep and Train ➤ Apache Spark, Azure Databricks, Apache Spark on HDInsight clusters

4. Model and Serve ➤ Cosmos DB, Synapse Analytics, Azure Data Explorer

All the services mentioned are managed services on Azure. This means, as a data engineer or administrator there is no overhead to patch, setup monitoring, high availability, or backups. All the operational activities are natively managed by Microsoft. However, all of this can be customized as per the requirements of the data analytics solution being built.

With this, we will mark the end of this chapter and the book.

Happy Learning!

# Index

## A

© Harsh Chawla and Pankaj Khattar 2020
H. Chawla and P. Khattar, *Data Lake Analytics on Microsoft Azure*,
https://doi.org/10.1007/978-1-4842-6252-8

Printed in the United States
By Bookmasters